FUNGI OF BELIZE

Russell J. Gray

Global Explorer Press

First edition printed 2018

Cover photos: [top] © Allen Rockefeller; [bottom left] © Gil Wizen; [bottom middle] © Craig Knox; [bottom right] © Ella Baron.
All other photos in this book are cited beneath the photo or in the Acknowledgements section of the book; photos without citation are either public domain/royalty free or property of Russell Gray.

Authoring, Layout Design & Formatting by Russell J. Gray

LCCN: 2018910368

ISBN13: 978-0-692-18355-7

Global Explorer Press
705 Alligator Ranch Rd.
New Smyrna Beach, Florida, 32168

Join the
Fungi of Belize
Facebook group and contribute to the growing knowledge of fungi in the nation!

TABLE OF CONTENTS

FRIENDLY REMINDER

Always remember: when mushroom hunting, if you have any doubts whatsoever about the identification of a fungus, do NOT eat it unless you have an expert opinion on ID. All information on edibility of the species described in this book was retrieved from field guides (Patterson's Field Guides: Mushrooms – Kent H. McKnight/Vera B. McKnight, Mushrooming Without Fear: The Beginner's guide to Collecting Safe and Delicious Mushrooms – Alexander Schwab, and Mushrooms of the Southeastern United States – Alan Bessette) in addition to various reputable online sources and peer-reviewed articles which are cited in the "References" section.

Methods used to identify fungi in this book included numerous field surveys, analyses, and crowd sourced input from mycologists and fungi enthusiasts. Most fungi species described in this book were analyzed for macroscopic and microscopic characteristics (some only macroscopic). Identifications were corroborated via field guides, iNaturalist community, Mushroom Observer community, Mushroom Identification page on Facebook, and the Mushroom Expert website. The various online communities assisted greatly in narrowing fungi down to the genus level, and sometimes to the species level when certain. The field guides and online descriptions assisted to verify these identifications. However, some rather obscure species required a more extensive literature review, and were named in this book after the closest species matching anatomical characteristics, proximity to known geographic distribution, habitat preference, substrate preference, and preferred environmental/climatic conditions. All sources used to compare and identify fungi are cited in the "References" section at the end of this book.

I am aware that species placement in mycology is ever-changing with new developments in the field. Some of the species identifications in this book may change or be inaccurate due to the complexities of mycological taxonomy and the lack of genetic sampling on fungi in Belize. Species names that are labeled in the ID keys should be treated as taxonomic placeholders and not as true species level identifications, as some may have microscopic or genetic differences that are yet to be studied and defined.

That being said, I insist that any inaccuracies or taxonomic rearrangements be brought to my attention so that future volumes of this guide can be updated. I can be contacted at **Russell_Graypt@Yahoo.com**.

ACKNOWLEDGEMENTS

In the making of this field guide, I have had several people assist me with various portions. I would like to credit each of them for their work and thank them for their incredible amount of help. This book would never be the same without them!

I would like to thank both Timothy J. Baroni and D. Jean Lodge, and their support for the work in Belize from the National Science Foundation, Biodiversity Surveys and Inventories Program to the State University of New York, College at Cortland (DEB-0103621). I was provided a multitude of baseline data by the work of D. Jean Lodge and Mr. Baroni, conducted under the Basidiomycetes of Neotropical Pine Forests Grant. These data allowed me to effectively compare my own information and ensure accuracy of identifications. Additionally, Dr. Jean Lodge assisted in proofing the identification key multiple times in order to ensure further accuracy, and also provided me with useful literature and experts to contact. I thank Dr. Jean Lodge again for her constant guidance with this book, regardless of her already-busy schedule.

My thanks go out to Alan Rockefeller who provided a significant amount of photos, and assistance with species identifications. Also a shout out to Counter Culture Labs, a public lab-space that has provided the necessary equipment to genetically sequence fungi he's found for accurate identifications.

I would like to acknowledge the iNaturalist and Mushroom Observer communities for assisting with direction on species identifications, experts to contact, and literature on fungi. These citizen science platforms have significantly reduced the time I had to personally spend researching, and for that I am grateful.

Finally, I would like to thank my mother, Leslie England, and father Timothy Gray for their continuous support of my projects, and encouragement to carry on.

Many residents of Belize have assisted me in my hunts for mushrooms, and many more folks on social media have assisted in providing photos and information. Some of the photos I used were modified and others are in the visual key without proper credit. On the following page, I would like to acknowledge each of them along with species images they provided, and corresponding page numbers.

Alan Rockefeller: Agaricus semotus (pg. 66), Amanita jacksonii (pg. 67), Amanita muscaria var. flavivolvata (pg. 67), Auricularia auricula-judae (pg. 84), Auriscalpium vulgare (pg. 19), bioluminescent mycena [right] (pg. 77), bioluminescent mushrooms [header] (pg. 34) Boletus dupainii (pg. 81), Cantharellus tubaeformis (pg. 68), Coprinus comatus (pg. 70), Lepiota cristata [gills] (pg. 74), Panaeolus antillarum (pg.78), Panaeolus cinctulus (pg. 78), Panaeolus cyanescens (pg. 78), Panaeolus papilionaceus (pg.79),Psylocybe cubensis [right] (pg. 80), Pulveroboletus ravenelii (pg. 82), Russula flavida (pg. 80)
Craig Knox: Cordyceps ant [right] (pg. 87; top & bottom 28), Phaeoclavulina cyanocephala (pg. 87), Pleurotus djamor [right] (pg. 65)
David Amaya: All 3 species of slime mold (pg. 91), Ganoderma sp. (pg. 59), Inonotus dryadeus (pg. 60), Stereum sp. (pg. 62)
D. Jean Lodge: Hygrocybe cf. occidentalis (pg. 73), Laetiporus caribeaus (right) (pg. 57), Laternea triscapa [right] (pg. 82), Mycena margarita [left] (pg. 77)
Eden Estefan: Aureboletus/Boletus auriporus (pg. 81), Boletellus ananas (pg. 81)
Ella Baron: Akanthomyces moth (pg. 87; 27), Coprinellus disseminates (pg. 70), Entoloma dragonosporum (pg. 71), Psathyrella hirta (pg. 80)
Gil Wizen: Phallus indusiatus (pg. 83)
Gillian Kirkwood: Gaestrumsaccatum [right] (pg. 90)
iNaturalist: iNaturalist. Available from http://www.inaturalist.org. Accessed [08/21/2018] (pg. 128)
Jacqueline Mitaritonno: Laetiporus sp. [left] (pg. 57)
Jerzy Opioła: Phaeolus schweinitzii [pores] (pg. 61)
John Plischke: Boletus variipes var. fagicola (pg, 82), Leccinum rugioceps (pg. 82), Lepiota cristata [left] (pg. 74), Pleurocybella porrigens[left] (pg. 65) Xylaria polymorpha (right) (pg. 88)
Kathleen Sullivan: Chlorophyllum molybdites (pg. 69)
Mandi Quark: Chemical bottles (pg. 134)
Marvin Paredes: Cyathus berkeleyanus (pg. 86), Ganoderma cf. resinaceum (pg. 59), Pycnoporus sanguineus[right] (pg. 61), Trametes hirsuta (pg. 63), Xylaria polymorpha [left] (pg. 88)
Maya Papovic: Clathyrus crispus (pg. 83)
Michelle Sevilla: All Mushroom Characteristic Illustrations (pg. 18, 20, 21, 22)
Ollie Meidinger: Leafcutter ant [bottom] (pg. 29)
Pat Spillane: Favolustenuiculus (pg. 58)
Paul Stevenson: Psilocybe cubensis [left] (pg. 80)
Tammy Trish: Photo of tube pores (pg. 19)
Thiago Mouzinho: Caripia montagnei (pg. 69)
Tom Astle: Marasmius haematocephalus (pg. 76)
Toucan Ridge Ecology and Education Society (T.R.E.E.S.): Cookeina tricholoma (pg. 86)

There are old mushroom hunters and bold mushroom hunters, but no old, bold mushroom hunters.

PREFACE

Fungi; a word that's proper pronunciation is under constant debate (is it "fun-jiy" or "fun-guy"?), refers to a kingdom of eukaryotic organisms that decompose organic matter, live symbiotically with plants and animals (or both), and act as parasites/pathogens. Fungi are not plants, nor are they animals; rather, they are somewhere in between. The kingdom of fungi is set in its particular phylogenetic placement primarily due to the presence of chitin in the cell wall (Bartnicki-Garcia, 1968). Chitin in fungi is a protective coating in the cells, the same structural compound that makes up the exoskeletons of arthropods. The presence of this compound pins the fungi kingdom closer to animals than to plants in regards to their evolutionary origins. Apart from being decomposers, fungi can be used for clinical/medicinal purposes (De Hoog, et al. 2000), biotechnology (Wainwright, 1992), cosmetic, and culinary purposes as well (Grieve 1971). Modern uses of fungus are still being tried and tested, though the future seems to be bright for the myriad of industries currently utilizing them.

Fungi come in a wide range of shapes and sizes. Some fungi can be as small as a solitary, single celled organism (e.g. yeasts and molds), and at the same time, they can be as larger than a blue whale. In fact, the honey mushroom (*Armillaria* sp.) spreading about 3.8 kilometers (2.4 miles) across the Blue Mountains of Oregon is the largest organism on the planet (Cassleman, 2007).

The following content of this book will be broken down into several sections to provide a well-rounded understanding of fungi in general, as well as the fungi found in Belize. Although the photo-identification section of this guide covers two of the major fungi phyla – Basidiomycota & Ascomycota – It is important to first review the other major fungi phyla as well. Knowledge of these divisions can answer the question "what can be considered a "mushroom" and what cannot".

Mycophobia (fear of mushrooms), plays a huge role in the lack of knowledge world-wide when it comes to mushroom hunting. It is my belief that this fear is greatly due to the large amount of poisonous species which can cause hallucinations, adverse effects on health, and even mortality when consumed. Due to these factors, it is important to people who wish to forage edible mushroom species to know the specific physical characteristics, substrates and habitats in which these fungi thrive, in addition to any similar species that should be avoided.

During my time in Belize, I noticed that there is a wealth of knowledge in regards to plants and animals held by both villagers and scholars alike. However, there seems to be very little knowledge about fungi. Even some of the most seasoned bushmen/woman wouldn't care to guess the identification of most mushrooms around. The reason for this, I believe, is simply lack of a tool that can assist with identification of common species. My wish is that upon the release of this book, there will be future generations of Belizean mushroom hunters that will enthusiastically hunt and identify various fungi, in order to utilize or study them as they please.

That being said, *Fungi of Belize* was created as a basic educational tool for finding and identifying wild mushrooms. This book is in no way an extremely comprehensive or technical manual that goes into tedious detail about every species described. I will not add intense scientific jargon to confuse the reader, and any technical terms I do use will be immediately explained in the text. The only thing I truly want to accomplish is to provide a simple means of understanding the fascinating fungi of the beautiful green jewel that is Belize.

Happy hunting to all!

AUTHOR'S NOTE:

A Mycological Message in a Bottle

I would like to begin this excerpt with a full disclosure: I am not a mycologist. My studies as a scientist are usually focused on reptiles and amphibians. Because of this, I have worked closely with professional mycologists and fungi experts in the making of this book to keep the information provided as accurate as possible. In addition, I've made my best effort to divulge and cite information throughout the text in hopes that readers will expand on the points that I make and continue researching.

Since this book is not aimed directly at the scientific community, I would like to briefly explain the APA citation format which is used. When you see a name followed by a date in parenthesis within the text – example: (Name, et al. 2018) – it is referring to a citation that can be found in the "References" section at the end of the book. If you are reading an article or excerpt within this book and you would like to learn more, I encourage you to seek out the referenced text and continue your research.

With all that out of the way, I'd like to introduce myself. Personally, I like to think of myself more as a naturalist; I admire all forms of nature. There is so much to learn from, so much to utilize, and thus, so much to protect. Among other things, the fungi kingdom has always been a fascination of mine. However, I never truly understood its complexity until I started intensively researching the cultivation process at about the age of 23. In short, the events that led to furthering my research of fungi, and ultimately the motivation that lead to writing this book are all derived from a single point in my life. Allow me to explain...

Growing up in the United States, I have always enjoyed mushrooms as an edible because they're readily available at restaurants, farmer's markets and grocery stores. I suppose it is because of these conveniences that I never really thought about their growth cycle, or what value they were to an ecosystem. It wasn't until I was working at a research station in Thailand that my fascination of fungi really took hold. I began to constantly notice their variety and beauty, stopping to take photos in the forests while I worked. I would see them at roadside stands and always buy one or two. Not even to eat, just to inspect and admire.

The research station was a known protected area, which made it illegal to remove any plants animals or fungi from the property for any reason. Unfortunately, many Thai locals from nearby villages would enter the protected area and poach the edible mushroom species during the rainy season due to their plentiful proliferation. However, when caught, they would be fined as a consequence of their actions. When I heard about this, a fellow researcher and I decided to try a few experiments to grow a simple fungus. The decided species was Oyster mushrooms (*Pleurotus ostreatus*) a beginners mushroom. We wanted to grow them from recycled materials: plastic bottles (container); spent coffee grounds, coconut husk, and old rice (substrate), using spore prints from the mushrooms. Our hope was that we could develop sustainable, low-tech mushroom farming practices and prevent further economic penalties for the local people and their families.

We had done our research (or so we thought), and we knew that our chosen substrates had worked for mushroom cultivation in the past; I had personally visited several mushroom farms that were using them. What we didn't know was that artificial mushroom cultivation was a very meticulous process which requires a completely sterile environment in its primary stages and relatively standard temperature and humidity in the later stages. Time and time again we tried; time and time again we failed. However, with each passing attempt, something new was learned; with each attempt some methodologies were improved.

Unfortunately, after a while I had left the field station, and was on my own. Though I was still determined to finish what I started. I began asking around and reading more into the cultivation process. I studied the ins and outs, the micro-details, the little bits and pieces that are picked only from the minds of mushroom farmers. These experiments were the birthplace of my passion and enthusiasm for mycology, and in the end, the experiment was a success…

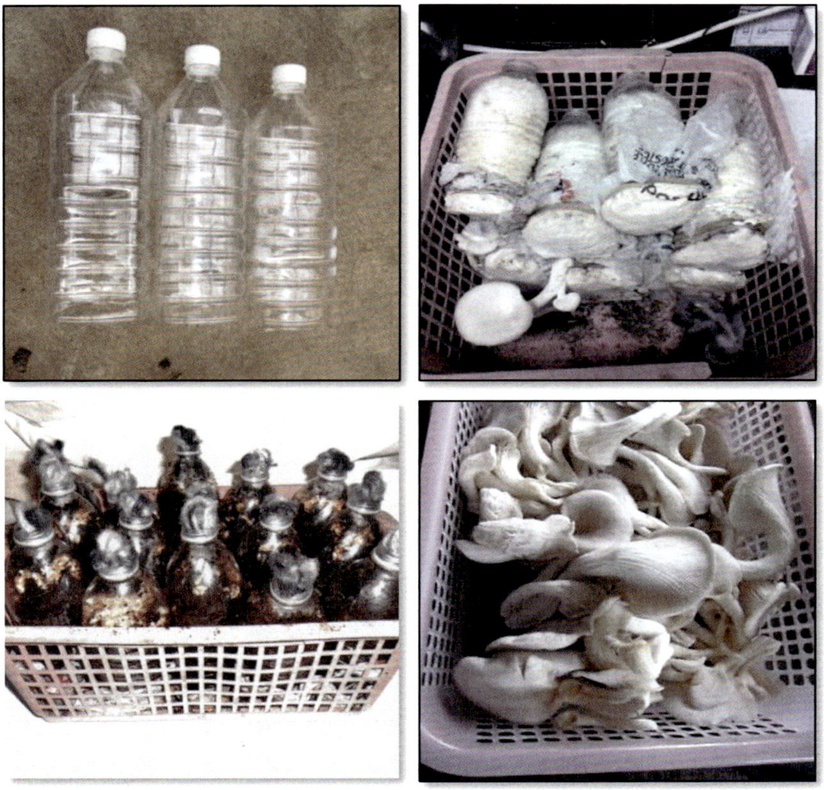

Artificial cultivation of oyster mushrooms (*Pleurotus ostreatus)* using recycled plastic bottles, plastic grocery bags, and spent coffee grounds collected from a coffee shop. A more in-depth explanation of artificial mushroom cultivation can be found on pages 42-51

Section I

KINGDOM: FUNGI

"Nature doth thus kindly heal every wound. By the mediation of a thousand little mosses and fungi, the most unsightly objects become radiant of beauty. There seem to be two sides of this world, presented us at different times, as we see things in growth or dissolution, in life or death."

-Henry David Thoreau

THE FIVE MAJOR FUNGI PHYLA

There are five major phyla of fungi that will be covered in this section in respects to their different characteristics, lifecycles and biological importance. These phyla include: Basidiomycota, Ascomycota, Zygomycota, Glomeromycota and Chytridiomycota. In addition to these classifications, this section will also cover the difference between fungi and lichen due to the common misconception that they are one in the same.

Basidiomycota (Club Fungi)

Fly Agaric (*Amanita muscaria var. flavivolvata*)
Photo by Alan Rockefeller

This phylum encompasses the macrofungi or "higher fungi" species within the fungi kingdom. Basidiomycota contains about 30,000 described species, 37% of which are "true fungi" (Kirk, et al. 2001). These fungi create vast systems of hyphal mycelium and give rise to fruiting bodies (e.g. mushrooms) which reproduce by sporulation. The sexual parts of these fungi are called basidia which bear external meiospores called basidiospores. These spores germinate in organic matter and sprout hyphae which branch out and multiply to create a mycelial system. The mycelium, which is the bulk of the organism, gives rise to spore producing fruiting bodies (Swann & Hibbett, 2007).

Ascomycota (Sac Fungi)

Venus' Wine Cup (*Cookeina speciosa)*

Sac fungi, or ascomycetes, encompass a phylum of over 64,000 species; making it the largest phylum of fungi (Kirk, et al. 2008).The most commonly known fungi of this phylum is the truffle, which is one of the most expensive fungi due to its culinary value and rarity. The most common macro-Ascomycete seen in Belize would likely be the Venus' Wine Cup (*Cookeina speciosa)*. The reproductive parts in these fungi are called ascospores, which are non-motile spores contained in the functional part of ascomycete fungi called the ascus. However, most species of ascomycetes can reproduce asexually too.

Zygomycota (Bread Molds)

Bread Mold (*Rhizopus stolonifer)*

A spore producing fungus that reproduces sexually. The phylum Zygomycota encompasses approximately 1050 species (Krogh, 2010); the most commonly known fungi of this phylum would be bread mold. Zygomycota grow via mycelial systems of thread-like, branching hyphae structures. The reproductive parts of Zygomycota are called zygophores; zygophores are chemotrophic hyphae that have attractive (-) and (+) variations which give rise to a zygospore. The zygospore germinates and grows into threadlike structural hyphae.

Glomeromycota (Arbuscular Mycorrhizal Fungi)

Ectomycorrhizal mycelium (white) associated to *Picea glauca* roots (brown)
Photo by André-Ph. D. Picard

Glomeromycota is a division of mycorrhizae which have a symbiotic relationship with the root systems, or thalli, of terrestrial plants by forming arbuscules (root penetrating hyphae) to help plants capture nutrients (Brundrett 2002). Within this division are approximately 230 described species (Schüßler & Walker, 2010). The lifecycle of glomeromycetes is very dependent on the root systems of plants; although their spores can germinate and grow in the absence of a host root, their hyphal growth is very limited.

Chytridiomycota (Chytrid Fungi)

Frog mortality by Chytrid fungus (*Batrachochytrium dendrobatidis*)
Photo by Jonathan Kolby

Chytridiomycota is the oldest known phyla of fungi, dating back to the Devonian period (Hass, et al. 1994). This phylum contains approximately 750 largely aquatic species with more being studied (James, et al. 2006). One very important organism from this phylum is the Chytrid fungus, which is currently threatening the world amphibian population with chytridiomycosis infection. Chytridiomycota fungi reproduce primarily through asexual means via zoospores; sexual reproduction in this phylum is rare (Hibbett, et al. 2007).

LICHEN VS. FUNGI

Another interesting variation of fungi is Lichen. These 'pseudo-fungi' are actually a mutualistic relationship between a fungus and a photosynthetic partner. There are three major types of lichen that you may even see on an everyday basis. Though keep in mind, lichens are not traditionally classified as fungi but rather their as their own organismal group.

Crustose

Crustose is a type of lichen which is characterized by the organism's close-binding relationship to its substrate (e.g. rocks and tree bark). Like all lichens, crustose is a fungal filament system in conjunction with algae. Crustose reproduces both sexually and asexually by dispersing macroconidia, then germinating and lichenizing the photobionts (green algae or cyano-bacteria cells). The reproduction process consists of ascospore dispersal allowing conidia and photobionts to colonize a substrate in conjunction with algae (Sanders, 2014).

Crustose lichen dappled on a tree

Foliose

Foliose is a type of lichen, characterized by its leaf-like shape. As mentioned before, lichens are a combination of both fungi and algae; this allows reproduction to occur both sexually and asexually through spores (sexual reproduction) or through vegetative growth (asexual reproduction) depending on the dominance of the algae or fungi growth (Easton, 1994).

Tufts of foliose lichen

Fruticose

Fruticose lichen is characterized by a bush-like structure composed of a root-like structure (holdfast) and an apically reaching thallus. Like other mentioned lichens, this organism is a combination of photobionts and mycobionts, however the photobionts are dominant (Nash 2010). Also, like previously mention lichens, fruticose can reproduce both by spores and vegetative growth.

Fruticose lichen emerging from moss
Photo by Michel Vorasane

ECONOMIC VALUES OF FUNGI

In order to exemplify economic significance of the five major phyla of fungi, each phylum can be analyzed to derive value of their associated members. Fungi are more important than many people realize!

Basidiomycota: This fungi group is comprised of the higher fungi, which are relatively large in size. Many of the fungi species encompassed in Basidiomycota are edible and therefore of economic value. Some, for example, *Pleurotus ostreatus* are not only edible, but can be used for bioremediation of crude oil spills (Isikhuemhen, Anoliefo & Oghale, 2003), which is of both economic and environmental value. Also, many species under this group can be used for medicinal purposes in teas and tinctures (e.g. *Trametes versicolor* and *Ganoderma lucidum*) and other psychedelic species, such as *Psilocybin cubensis* are being investigated for their ability to combat psychological disorders, like depression (Grob et al. 2011).

Ascomycota: This fungi group is comprised of sac fungi, leaf fungus, yeasts, truffles and plant pathogens. One of the most commonly known organisms under this group is *Penicillium*, which is a major component of the antibiotic drug penicillin; this alone is of great medical and economic value. Other members of this group such as truffles and yeasts can be used for culinary purposes; truffles are one of the most expensive edible fungi.

Zygomycota: These fungi have several uses that signify their economic importance such as culinary (Indian tempeh and Chinese sufu), used as components of birth control pills and also as dyes or food pigmentation (e.g. the yellow coloration of margarine).

Glomeromycota: The mycorrhizal fungi that make up this phylum are a very important tool in agriculture nowadays; especially soybean, cassava and coffee. When the fungus attaches itself to the roots of crops it creates a network between the plants and, in a sense, communicates between them. When one plant is deficient in a certain nutrient and another in the network has enough to spare, the fungus can obtain and transfer that nutrient between plants. Additionally, mycorrhizae can improve the immune system of plants that they have intertwined with; this phenomenon has been termed "priming". There are several interesting studies and presentations outlined in the BBC article: "Plants Talk to Each Other Using an Internet of Fungus" which explain these fungi and how plants in a forested area can "talk to each other" through mycorrhizal fungi systems (Flemming, 2014)

Chytridiomycota: Although the organisms in this group are not known to have any positive economic value, they are of a negative economic importance as a large number of the species are parasitic to plants and animals.

Wine, cheese, and bread all require yeast fungi to be produced.

LIFECYCLE OF FUNGI

To begin this section, let's go over the basics of the mushroom lifecycle from "the ground up", so to speak, and identify some of the key terms when talking about mushrooms and fungi…

In the process of decomposing organic matter, fungal growth often starts with spores. Spores are very small "seeds" of fungi and are the initial stage of fungal growth. When spores are dropped on an ideal substrate (i.e. wood, straw, coconut husk or other organic matter) in ideal conditions (humid, cool weather in a relatively sun-free area) they germinate into what is called hyphae. Fungal hyphae are the cellular building block of fungi that connect with one another to form the primary part of the organism – the mycelium. You may have seen mycelial growth out in the forest while hiking and didn't even know or recognize it! The pattern sometimes appears like white branching webs that can be easily seen creeping and spreading over fallen leaves or rotting wood. The mycelium secretes enzymes on its substrate and expands its mycelial network until it engulfs the substrate and has the energy requirements needed to produce fruiting bodies. When a fruiting body grows and matures, it releases spores to be either carried by wind or animals to the next substrate, or produces root-like structures to locate and colonize the next substrate; thus, continuing the process of decomposition. The mycelial network makes up the majority of the organism; the fruiting body is simply the means of reproduction.

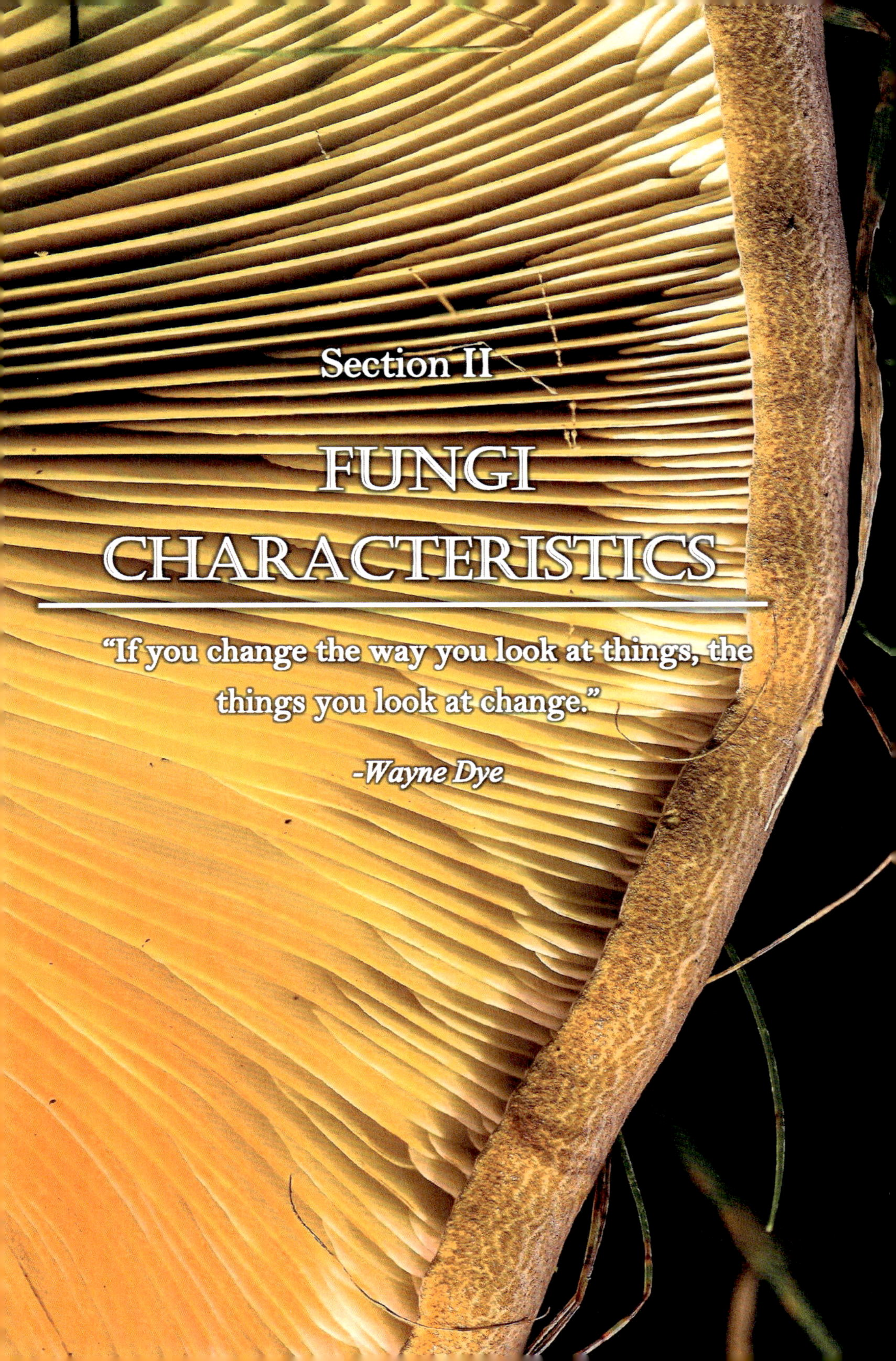

Section II

FUNGI CHARACTERISTICS

"If you change the way you look at things, the things you look at change."

-Wayne Dye

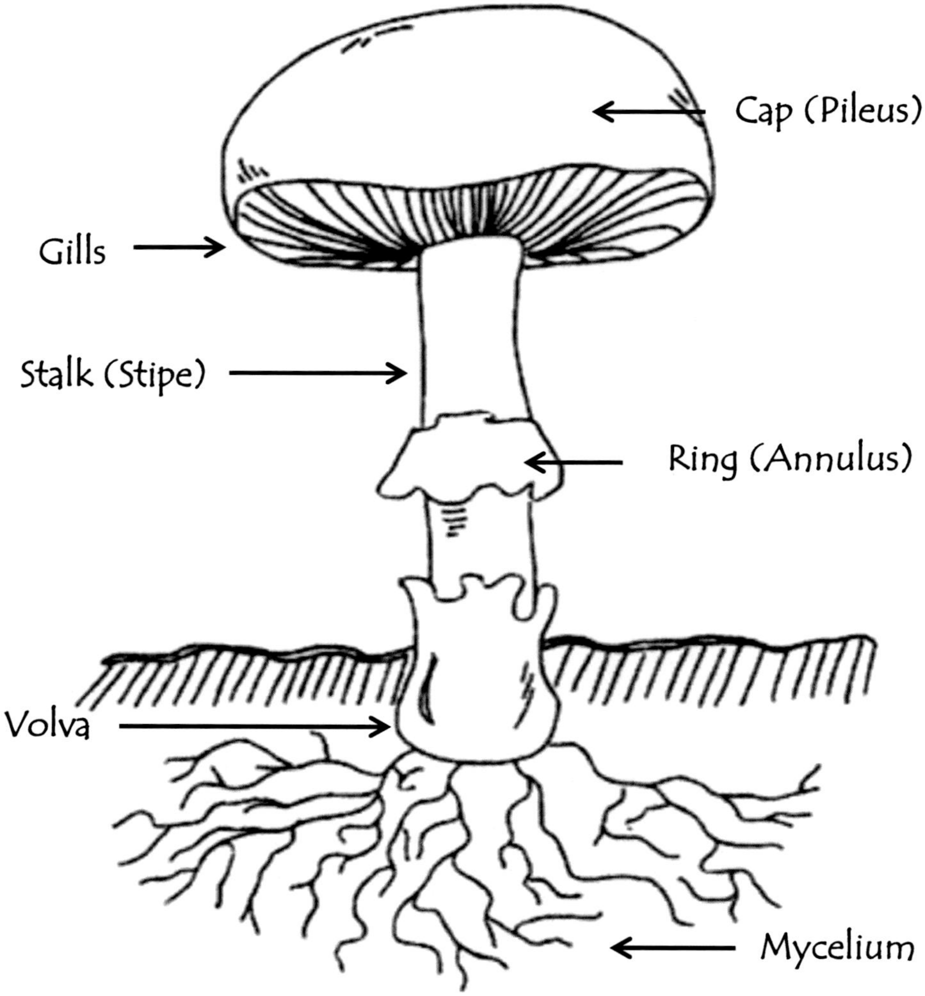
Cap (Pileus)
Gills
Stalk (Stipe)
Ring (Annulus)
Volva
Mycelium

Mushrooms can have several different spore-releasing parts that are found underneath the cap. These features are very helpful in the identification process, especially when trying to find edibles.

There are several physical characteristics to look out for when attempting to identify mushrooms. This section will provide you with basic illustrated examples, as well as the respective terminologies that will be used later on in the photo-identification section.

Under The Cap

Gills
Go from the cap to the stalk; brittle and easy to break off when rubbed

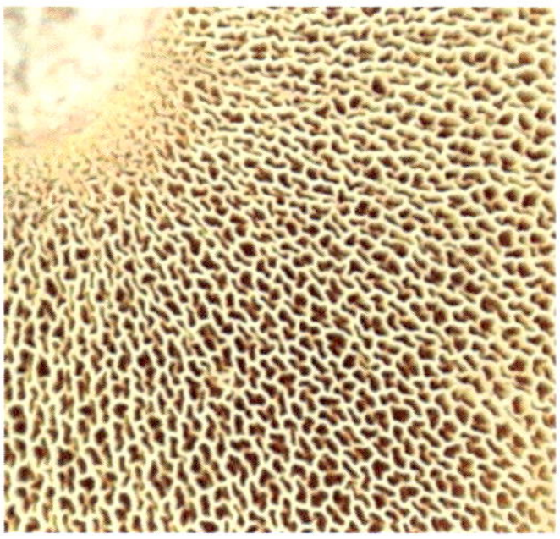

Tubes
Sponge-like feature in place of gills (typically found on boletes)

Ridges
Thin, shallow gills that are fixed to the cap and stalk

Pores
Many small holes; typically on bracket fungi

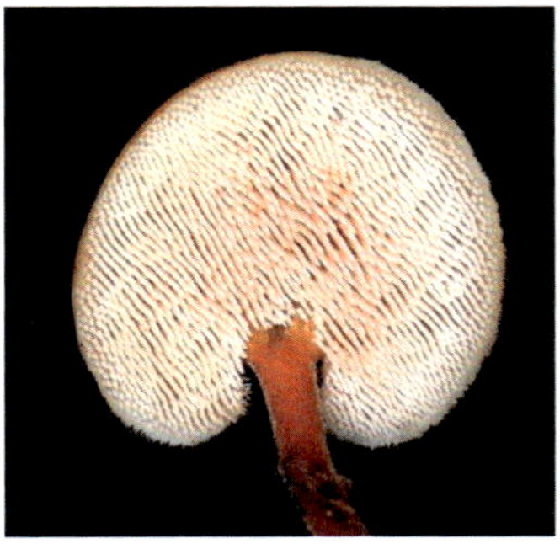

Spines/Teeth
Spikey stalactite-looking features on the underside of the cap

None
Some fungi, like ascomycetes, have no gill or pore-like features

Cap Textures

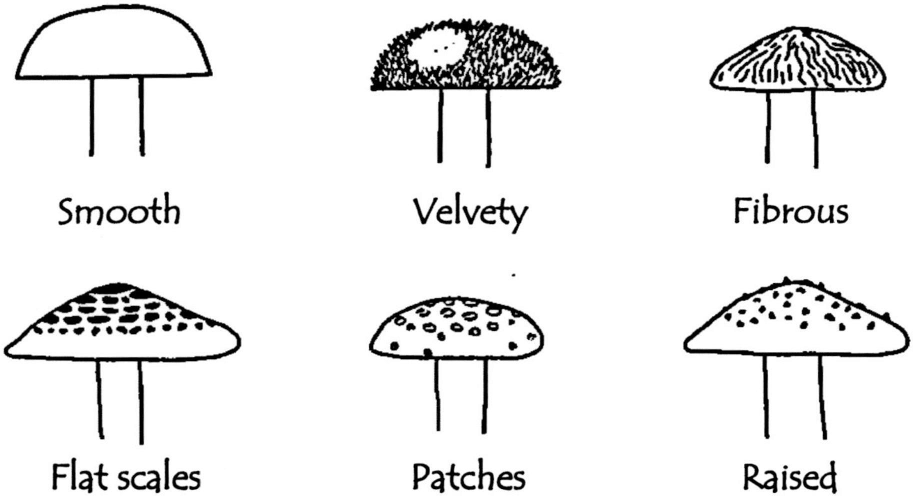

Gill Spacing

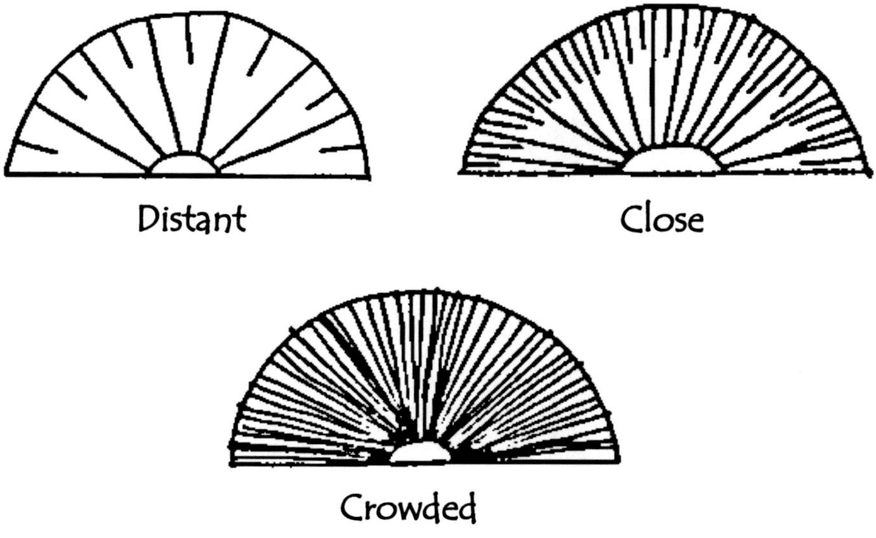

Cap Margins

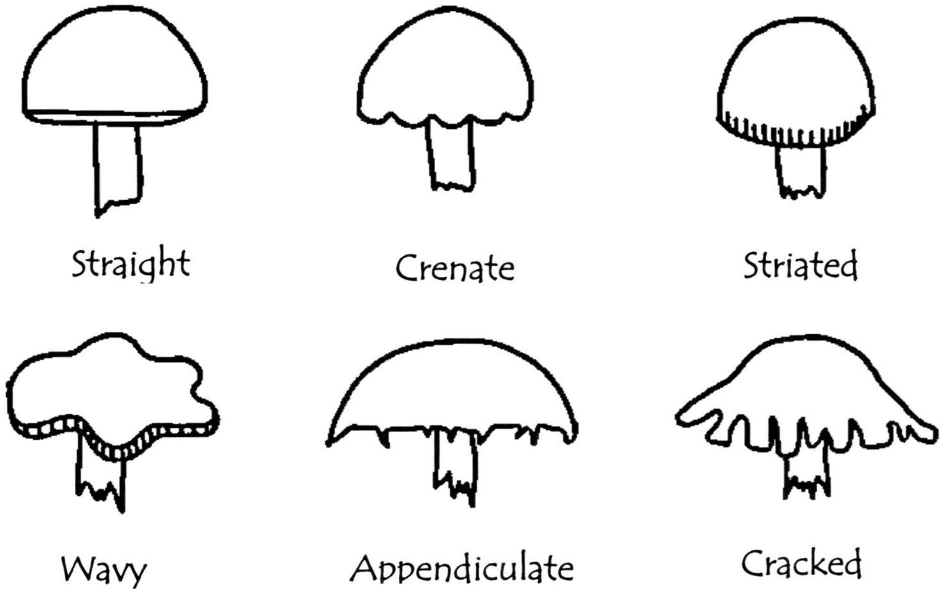

Gill Attachments

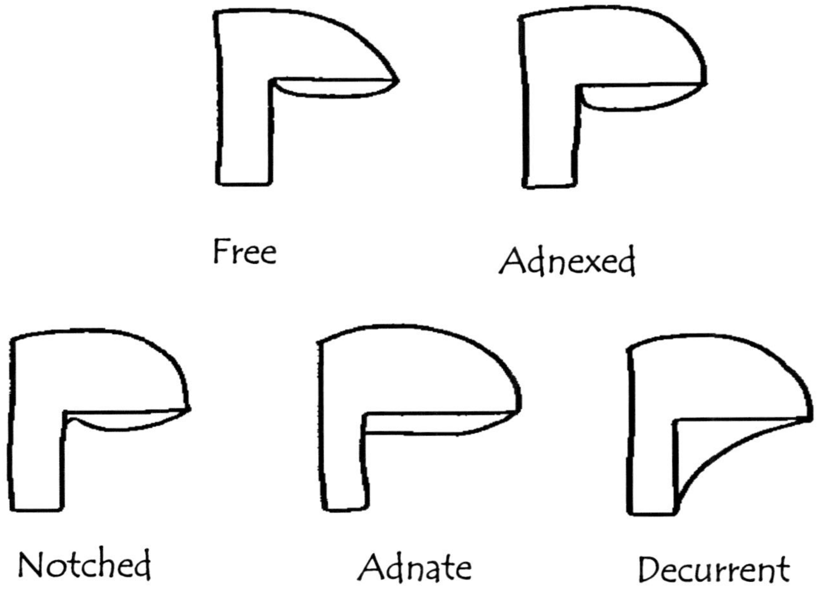

Cap Shape

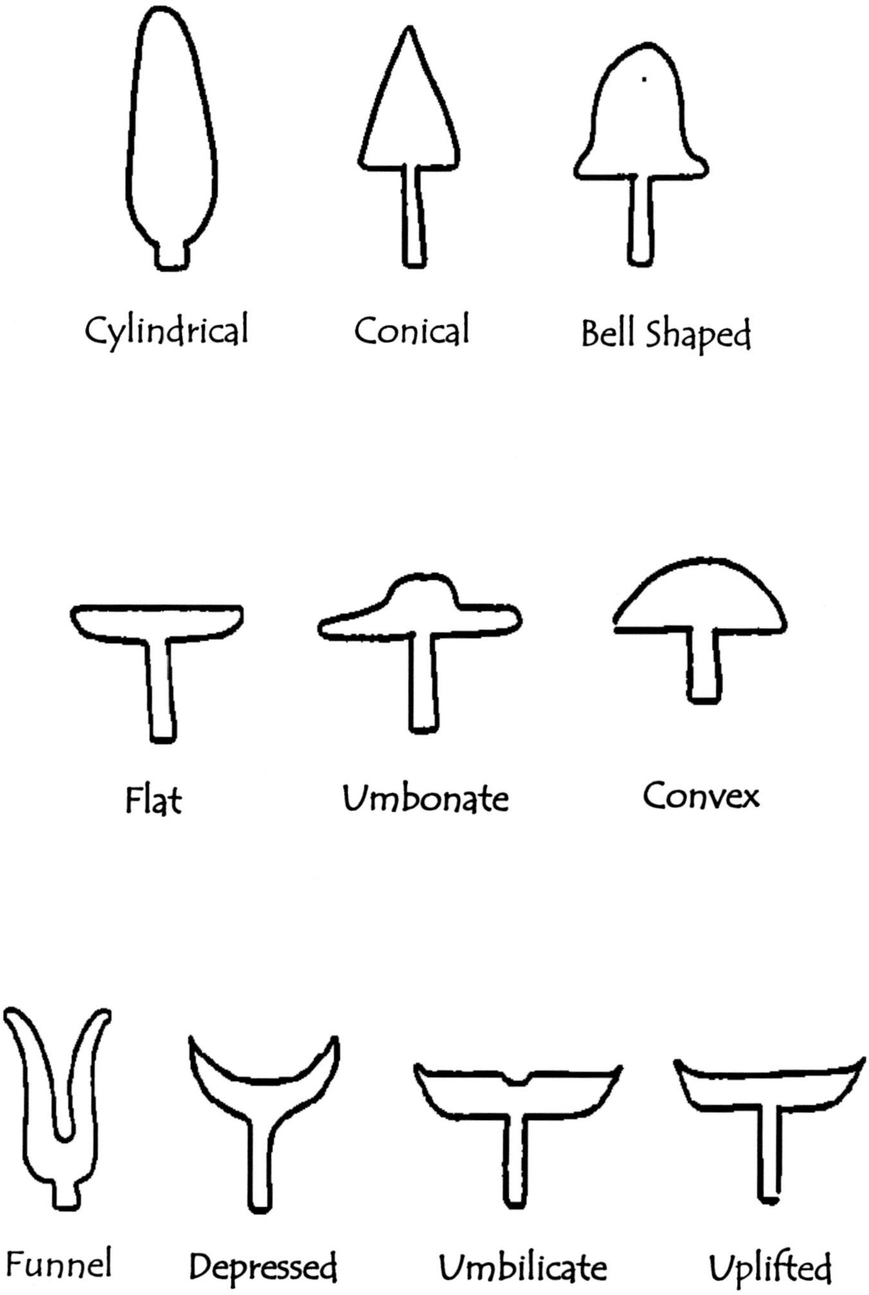

Cylindrical
Conical
Bell Shaped
Flat
Umbonate
Convex
Funnel
Depressed
Umbilicate
Uplifted

Section III

NATURAL HISTORY OF FUNGI IN BELIZE

"Fungi are the interface organisms between life and death"

-Paul Stamets

MUSHROOMS & THE MAYA

The word for mushroom in Yucatec Maya language or 'Màaya t'àan' is "Kuuxum, XuxumChé, & XikinChe'"

Ancient Maya civilizations spanned across most of Central America. The remnants of their cities and temples can be seen even today in countries like Belize. Interestingly enough, among the various artifacts and inscriptions that archaeologists have uncovered from the ancient world, are strange mushroom shaped idols. There is still some controversy and contention over what these idols represent; however, archaeologists believe they were a pestle used in the grinding of hallucinogenic mushrooms for ceremonial use (De Borhegyi, 1961).

Various Mushroom Stones (approx 1 ft tall - 1000 B.C. to 500 A.D.)

Various Mushroom Stones (approx. 1 ft. tall – 1000 B.C. to 500 A.D.

Photographer unknown; public domain

The use of hallucinogenic substances is widespread in ancient cultures. In the northern reaches of Mexico and borderlands of Texas, the Huichol tribes would use peyote (*Lophophora williamsii*) for their visionary sacraments (Schaefer & Furst, 1997). Vision quests induced by the powerful hallucinogenic plant were considered a ritualistic necessity to access the spirit world. The Native American Church still practices religious rites with peyote and have even deemed their religion *Peyotism*.

Another hallucinogenic plant that has been used since ancient times goes by the name *Ayahuasca*. Vines from the *Banisteriopsis caapi* plant, native to the Peruvian Amazon, are collected by indigenous peoples to create a potent psychoactive brew. Shamans use ayahuasca as a sort of medicine, though not in a traditional sense of cosmetic damage or physical illness. This medicine is meant for the soul.

With all this information considered, it doesn't seem too far-fetched an idea that Maya culture would embrace the practicesof ritualistic psychoactive substance use to alter their state of consciousness. There is an abundance of hallucinogenic species that occur in the Mesoamerican region, namely species from the genera *Panaeolus* and *Psilocybe.* It is suspected that the ancient cultures who resided in Mexico, Guatemala, and Belize targeted the species *Psilocybe mexicana* for their ritual ceremonies. In the early writings of Shultes and Hoffman (1980), it was reported that while reviewing texts from the early 1800's there was talk of the Mexican natives consuming small brown mushrooms they referred to as *Teonanácatl*. The word teonanácatl was of the Azteca language, which translates to "sacred mushroom" or "intoxicating mushroom", depending on the source (Badham, 1984; Singer, 1958).

Nowadays, psychoactive mushrooms are outlawed in most countries and frowned upon in most cultures. In a fascinating turn of historical events, something that was once revered as sacred is now considered profane.

ZOMBIES OF THE FOREST

Yes, you read the title right! Zombies are real (kind of), and they live in Belize!

With all sensationalism aside, there are some disturbing, yet intriguing pathogenic organisms which should never be left out of a good fungi book. What I'm referring to, of course, is Cordycepioid fungi (species within the families Cordycipitaceae, Ophiocordycipitaceae and Clavicipitaceae). Some cordycepioid fungi infect arthropods by taking control of their body, and in a sense, turning them into

FUNGAL ZOMBIES!

Although sometimes difficult to see, these pathogenic fungi can be found throughout the moist forested areas of Belize. Typically *Ophicordyceps* is found proximal to large ant colonies, which can be infected and enable the fungus to reproduce. Most cordycepioid fungi have specific hosts. The *Ophiocordyceps unilateralis* group, for example, usually targets *Camponotus* (carpenter ant) populations.

The fungus enters its ant host by means of tiny spores; these spores attach themselves to the cuticle of the ant while it's foraging for food, and upon entry into the exoskeleton, begins germinating (Hughes, et al. 2011). Once the hyphae (branching components of mycelium) have connected throughout the body, the fungus literally takes control of the ant's brain and manipulates the limbs to climb branches and leaves. When the fungus has reached an area where conditions are optimal to form a fruiting body, it forces the ant to clamp down its mandibles and latch itself to the leaf or branch. After a short period, the fruiting body emerges from its head, killing the ant and releasing more spores; Thus, continuing the lifecycle of the fungus.

Cordycepioid fungus can affect a range of arthropod hosts including spiders, grasshoppers, moths, and moth larvae. For example, in some Asian countries, people farm the cordyceps species *Ophiocordyceps sinensis* using the larvae of Ghost Moths (*Hepialus humuli*), the product of which is called "the golden worm". The fungi infected larvae sells as an herbal remedy and health supplement. It is believed that the golden worm can help to treat diseases related to the heart, lungs and kidneys; some say it also treats erectile dysfunction and even cancer. There have been somewhat promising studies done on mice using this strain of cordyceps fungus to test its chemical and medicinal components; however, there has been no significant research based evidence that proves its medicinal values in humans.

ANT FARM

Leafcutter ants are possibly the first farmers in history!

Usually, when someone talks about an "ant farm", they're referring to a glass container in which ant colony can be observed in the works. However, this is not at all what is meant when talking about the leaf-cutter ants of Belize. Many people tend to see the leafcutters marching to and fro with mandibles locked and leaves overhead. However, they fail to realize the intriguing agricultural event which takes place within the confines of a colony. Deep within the chambers of the leafcutter mound, is an actual farm being run entirely by ants!

In Belize, leaf-cutter ants (*Atta cephalotes*) cut leaves and make extensive 'highways' to transport them back to their nesting grounds. When inside the mound, there are vast tunnel systems and chambers which all have different purposes. One of these chambers is inoculated with fungus from the *Lepiotaceae* family. The fungus is cultivated by the decomposing of the leaves that are brought into the chamber and then used to feed the colony. Due to this constant interaction between ant and fungi, many leafcutter ants have grown resistant to many fungal pathogens, including the *Ophiordyceps* fungus previously mentioned in the book (Hughes, et al. 2003). Although extremely rare,

some leafcutters still get infected with cordycepioid fungi. Maybe those ones aren't very good farmers!

It is well documented that ants perform tasks in highly organized social structures and show signs of intelligence in their focused, systematic operations. So it isn't illogical to think they have left the "hunter/gatherer" stage of social/biological evolution, and began the "agrarian" stage, just as humans have in the past. Or is it the other way around? According to Mueller, et al. (1998), ants have been farming for to sustain their colonies millions of years before humans!

Leafcutter ants (*Atta texana*) in the fungal chamber of their colony
Photo by Alex Wild; Public Domain

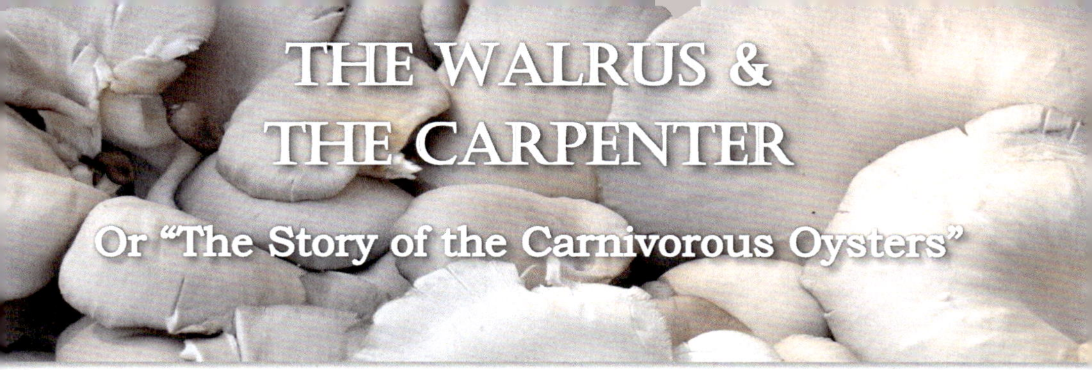

THE WALRUS & THE CARPENTER

Or "The Story of the Carnivorous Oysters"

If you're anything like me, you love Alice in Wonderland; especially the old 1951 cartoon version. So what, you may ask, does this amazingly whimsical cartoon movie have to do with fungi? Well. Other than the numerous fungi depicted in the movie, there is a scene where a story is explained called "The story of the curious oysters". In the story, the oysters all get eaten at the end due to their curiosity. That being said, the oysters I will be talking about in this article are curious too; however, they do the eating!

While researching mycelial growth, scientists have discovered that a few species of Oyster mushrooms including *Pleurotus ostreatus* and *Hohenbuehelia* sp. exhibit a unique behavior during the growth and expansion of their mycelium. The fungi attacks, captures and consumes nematodes and other microorganisms!

According to an article on this phenomenon by George L. Barron, titled "Predatory Fungi, Wood Decay, and the Carbon Cycle" (2011), this fungal predator/prey relationship occurs most often with nematodes, which are very small (100-1000 microns in length), non-segmented worms from the phylum Nematoda. The vast majority of these microorganisms feed on bacteria, fungi, and spores; because of this, they are very abundant in soil. During the study, it was found that species of *Nematoconus* and *Hohenbuehelia* begin to grow "hourglass-shaped" adhesive cells as extensions from the hyphae when in the presence of nematodes. Interestingly enough, these hour-glass projections are used as a lasso-like tool to capture their nematode prey and consume it. On the other hand, Oyster mushrooms (*Pleurotus ostreatus)* appear to release a toxin that paralyses nematodes, which are then targeted by specialized fungal cells that penetrate and enable consumption. It's kind of like a spider draining the life from its prey!

So why does this happen? What purpose would there be for a detritivorous organism to feed on prey that is still alive? Well, the current hypothesis is that the function of predation on nematodes may assist with intake of nitrogen compounds. Since these species are wood decay fungi, the nitrogen compounds obtained from the consumption of nematodes may help in the process of wood biodegradation. This theory has also concluded that due to trends in evolutionary history, it is likely that, in time, some fungi may diverge from the norm and take on the niche of exclusively predating nematodes. A true predatory fungus!

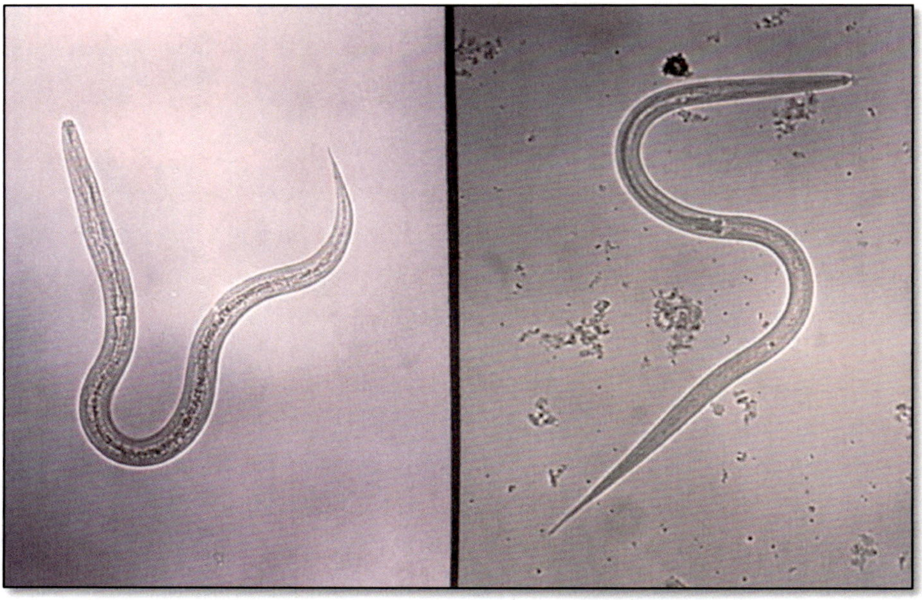

Hookworm nematodes (*Strongyloides* sp.) under microscopic magnification
Photo by CDC/Dr. Mae Melvin

BIOLUMINESCENT FUNGI

Turn off your flashlights in the forest at night and look for a delicate green glow

Glowing mushrooms in mythology and folklore were referred to as "foxfire". This mysterious green light of the forest used to be regarded as an occurrence of magic. We know today that foxfire isn't really magic, but instead a chemical reaction within the fungus causing an irradiant glow.

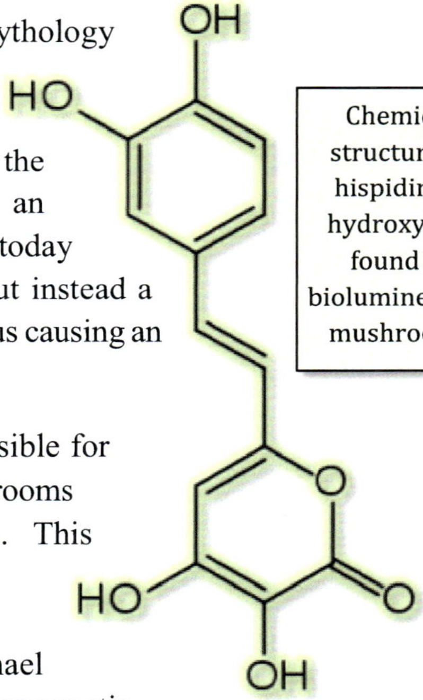

Chemical structure of hispidin-3-hydroxylase found in bioluminescent mushrooms

The chemical process responsible for bioluminescent reactions in mushrooms involves hispidin-3-hydroxylase. This compound is an enzymatic converter that creates what was first considered "luciferin" by Raphael Dubois in the 19th century, and its enzymatic partner, luciferase (Poisson, 2010). Luciferase catalyzes the oxidative reaction for luciferin and in doing this, produces energy in the form of light (Purtov, et al. 2015). Although the true purpose of this glowing biological function is unknown, the current consensus follows the theory that the light may attract nocturnal animals which assist the fungus in spore dispersal.

Another hypothesis for bioluminescence in fungi is that the light functions as a warning of toxicity. It is possible that both hypotheses could be true depending on the various genera/species of fungi.

Bioluminescent mycelium on a leaf found in Chignautla, Puebla, Mexico
Photo by Alan Rockefeller

Accounts of glowing mushroom species have been reported from all around the country of Belize. While walking through the forests on moonless nights in the rainy season, some say they have seen a dim blue-green glow emanating from the Maya Mountains. Others, to their surprise, have stumbled upon these majestic fungi on the rotting wood and seeds of old and dying cohune palms (Dourson, 2013).

Of 71 known bioluminescent (glowing) fungi species, the majority are part of the genus *Mycena*. According to the article "Luminescent Mycena: New and Noteworthy Species" by Desjadin, et al. (2010), seven new species of bioluminescent *Mycena* (4 new species and 3 previously described species) were found. One of the known species in Belize is called *Mycena margarita* (Alves & Nascimento, 2014).

Bioluminescent *Mycena* sp. from San Pedro Yolox, Oaxaca, Mexico
Photo by Alan Rockefeller

TROUBLE IN MOUNTAIN PINE RIDGE

One of the more unique ecosystems in Belize is known as Mountain Pine Ridge (MPR). This distinct location is now considered a pine forest reserve, which was established in 1944. MPR covers approximately 430 square kilometers and is located in the Cayo district, South of San Ignacio. The MPR ecoregion consists largely of montane pine forests and open, scattered grass-lands. Fragments of riparian broad-leaf forest and mixed deciduous ecotone can be found in its lower altitude areas. Several tributaries run through MPR and flow into the Macal River located in its southern extent.

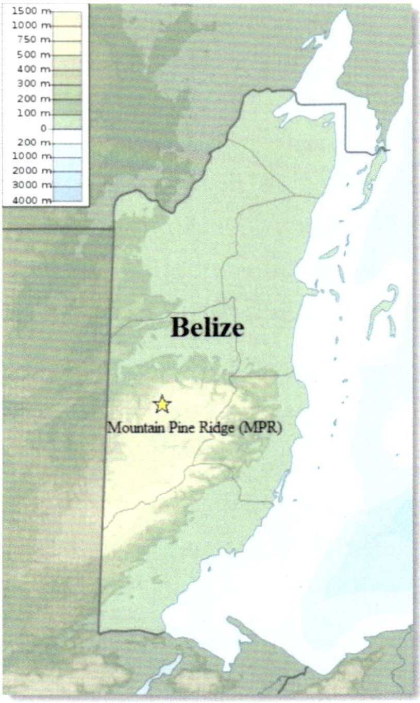

Location of Mountain Pine Ridge, Belize

A paper which was submitted to the XII World Forestry Congress by Hodgson & Darrow (2003) reported that in the early 2000s, over 90% of the MPR forest area was decimated by voracious bark beetles. Sadly, what was once a lush mountain pine forest has since been reduced to a grassy savannah with scattered pine trees. To make matters worse, fires and insufficient sapling growth may prevent MPR from recovering to its former productivity in our lifetime.

Southern pine beetle
(Dendroctonus frontalis)
Erich G. Vallery, USDA Forest Service - SRS-4552, Bugwood.org

The dominant species of tree in MPR is the Honduras pine (*Pinus hondurensi)*. A tree which has great appeal to several species of tropical bark dwelling beetles of genera *Dendroctonus* and *Ips*. The bark beetles bore into trees and cultivate ophiostomatoid wood-stain fungus in order to prevent the tree's defensive sap release. According to studies by Xudong, et al. (2004), and Moreno-Rico, et al. (2015), the species of bark beetle *Dendroctonus mexicanus* and *Ips calligraphus* are associated with ophiostomatoid fungi from the genera *Ophiostoma, Ceratocystiopsis*, and *Leptographium.* Once the sap has been neutralized by the fungus, an ideal environment is established for the beetles to lay eggs and reproduce. Some bark beetles, such as *Dendroctonus frontalis* deploy fungus on the exterior of the trees without burrowing; consequently, this species is responsible for the loss of pine forests across Central America (Billings, et al. 2004).

Bark beetles and their symbiotic fungus partners have proved to be devastating to the pine tree population of MPR and many other pine forests throughout North America, namely the western regional United States and Canada. The function and power of the fungi kingdom is not to be underestimated. Whether it be *Ophiocordyceps*, controlling ants from the inside to reproduce, or wood staining fungus using bark beetles to bring down entire forests, these otherwise stationary organisms seem to travel at will!

The road to Baldy Beacon, Mountain Pine Ridge, Belize

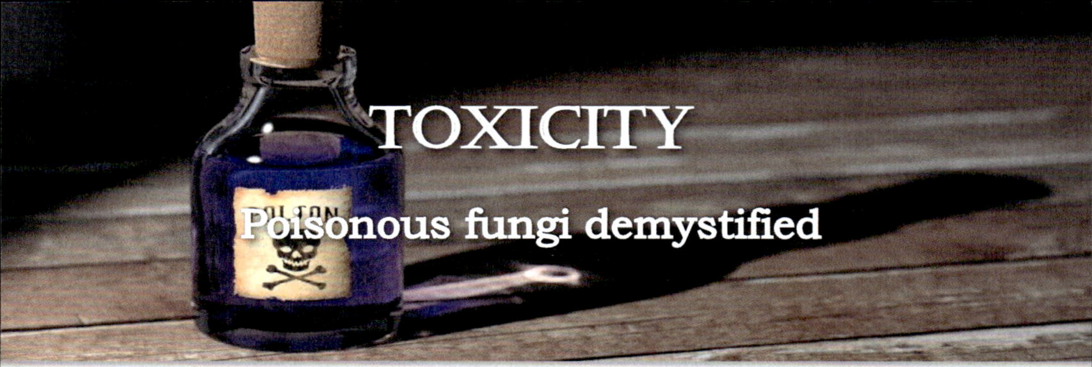

TOXICITY

Poisonous fungi demystified

Deathcap: A name given to *Amanita phalloides*; the deadliest species of mushroom on earth. Interestingly enough, some of the fortunate few to have mistakenly eat it and live say it was *the most pleasant tasting mushroom they'd ever eaten*.

Deathcap mushrooms (*Amanita phalloides*)
Photo by Alan Rockefeller

So what happens when you eat a severely poisonous mushroom like a deathcap? Well, symptoms generally start off with gastrointestinal distress, followed by waves of nausea and severe abdominal pain, followed by death. After about a day, the nausea and pain may subside for some, which misleads the victim to think they're getting better. However, in reality, their livers have been devastated. Those who don't die from eating the deathcap usually require liver transplants in order to survive. Although the deathcap hasn't been reported in Belize, other deadly mushroom species, such as the Deadly Gallerina (*Gallerina marginata*) and the Destroying Angel (*Amanita bisporigera*) have been documented. That being said, don't let this information discourage you from mushroom hunting. If anything, use these toxic mushrooms as examples of why you should learn more!

Now, with the worst case scenario out of the way, let's talk about the realities of biotoxins, or rather what they can and cannot do. The word *biotoxin* refers to a toxic substance that originates from a living organism. There are phytotoxins (toxins from plants), zootoxins (toxins from animals) and mycotoxins (toxins from fungi). In order to understand the poison in mushrooms, you must first understand the toxins in other living things.

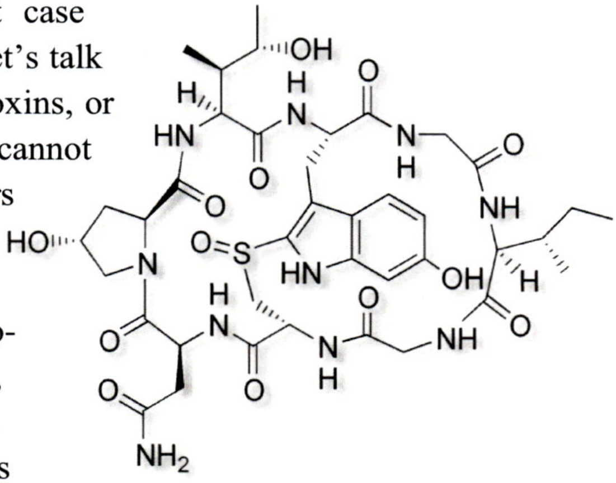

Structure of the oligopeptide γ-*Amanitin* ($C_{39}H_{54}N_{10}O_{13}S$); a primary amatoxin found in the deathcap mushroom

To elaborate, consider the properties of phytotoxins found in cocoyam (*Colocasia esculenta*). When cocoyam is eaten raw, it is poisonous. However, if you cook the plant, it becomes edible. This is because some plants have poisonous chemical compound and proteins that can be denatured or neutralized with heat. In the case of cocoyam, calcium carbonate is the chemical being removed. Unlike phtytoxins in cocoyam, most mycotoxins cannot be destroyed by heat. For example, amatoxin, the oligopeptide found in deathcap mushrooms is heat stable at roughly 250–280 °C (Wieland, et al. 1978). The average temperature that we cook food is about 75°C. All that aside, it is generally necessary to cook wild mushrooms, even edible species. Not because they are truly poisonous when raw, but because they can bioaccumulate toxins from the substrate on which they grow (with a few exceptions).

Now, consider the properties of zootoxins. Usually, the first thing that comes to mind when thinking about toxic animals is venom. Venoms are zootoxic substances which are delivered by mechanical injury (i.e. bites and stings). This form of toxin is made of proteins that not only denature in heat, but can also be denature in stomach acids. That's right, you can swallow venom and nothing would happen! However, if you have any cuts in your mouth, it can enter the blood stream, so don't try. On the other hand, there are also poisonous zootoxins. For example, poison dart frogs can be toxic through entry into the bloodstream and through digestion; their toxins may even be absorbed through the skin. To explain, dart frogs (e.g. *Phyllobates*sp.) contain tiny steroidal alkaloid compounds which can easily pass through plasma membranes (Bosmans, et al. 2004). Since these chemical compounds are so small, they have the ability to be dynamic in their toxicity.

Mexican West-Coast Rattlesnake (*Crotalusbasiliscus*) resting near Yolk Colored Caesar Mushrooms (*Amanita yema*)
Photo by Alan Rockefeller

In contrast to animal and plant toxins, the oligopeptide chains that comprise the poisons in mushrooms are too large to be absorbed through skin, and don't have the active binding sites to affect the body via bloodstream. Because of this, mushrooms are exclusively poisonous, and only toxic through ingestion. However, their poisons cannot be denatured or removed by heat. Although you cannot be poisoned by touching any mushrooms, there have been reported cases of contact dermatitis and other allergic reactions. However, the most likely people to become allergic are those who are constantly exposed in their occupations (Curnow & Tam, 2003; Liebeskind, 1972).

MARINE FUNGI

Oceanic decomposers

Believe it or not, there are fungi that live in oceans and estuaries; these fungi are considered "Marine Fungi". According to the archives of the University of Mississippi (2013), there are approximately 444 known species of marine fungi which are encompassed by the phyla Basidiomycota, Ascomycota, and Zygomycota. There are estimated to be about 1500 species worldwide (Hyde, et al. 1998). Marine fungi, like all other fungi, are decomposers; however, they have a variety of functions and specializations that separate their ecological roles from exclusively terrestrial fungi.

Fungi play important roles in all ecosystems, as pathogens, detritivorous organisms, and beneficial symbionts. Marine fungi decompose a wide range of organic matter and marine organisms, including corals, fish, marine plants, and algae (Kohlmeyer & Kohlmeyer, 2013). Some species are even known to penetrate and decompose calcareous substances like the shells of marine mollusks, as well as calcified algae (Kohlmeyer, 1969a).

Other than the presence of hosts and substrates, habitat choices of marine fungi have very little restriction. Species of marine fungi have been documented in regions of tropical, temperate, cosmopolitan, and even arctic climate conditions (Jones, 2000). Although marine fungi appear to have a wide distribution, they are more abundant in tropical and temperate regions, and their survival is measurably affected by the temperature of their environment (Boyd & Kohlmeyer, 1982; Kohlmeyer, 1983). It also appears that most species of marine fungi are constrained by depths of water in which they can survive, due to low oxygen levels

(Kohlmeyer, 1969b). Besides this, salinity itself doesn't seem to be a significant limiting factor (Byrne & Jones, 1975; Masuma, et al. 2001).

In Belize, you may encounter a number of tropical marine fungi without even realizing it. A study by Kohlmeyer (1984) documented 36 species, concluding that Belize was more biodiverse in marine fungi than any other site studied. Mangrove trees (*Rhizophora* spp.), for example, attract a number of marine fungi taxa. The trees shed their leaves fairly often, and heavy winds break their branches. Without marine fungi, the biomass from these mangrove forests would likely pile up. Although other detritivorous organisms do occur in the waters of Belize, marine fungi play an important role in softening the outer layers of the wood; thus enabling other marine animals, such as shipworms, to bore in, settle and reproduce (Kohlmeyer, et al. 1995). Besides the waters, tropical marine fungi are found on beaches too, sometimes protruding from the sands while parasitizing the roots of nearby trees.

On a more personal note, I find it truly incredible how many significant organisms may be overlooked in our exploration of the earth. Marine fungi are a 'textbook example', so to speak, considering approximately 70% of our planet consists of ocean. There are so many unanswered questions, and so many potential values that these organisms could contribute to our lives. With proper study, marine fungi could become our future medicines, or means to improve environmental mitigation efforts. We just have to be more curious, educated, and involved.

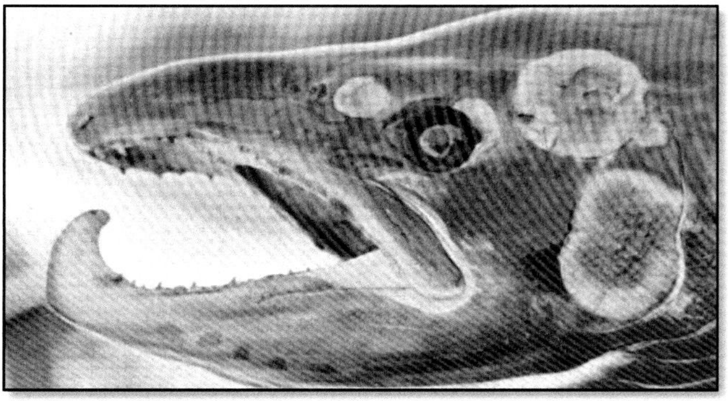

An example of marine fungi: 7-lb. Male Grilse, showing fungus partly developed on head. December 1905. Photo by Malloch, P. D. Public Domain

Section IV

MUSHROOM CULTIVATION

Big things often have small beginnings

ARTIFICIAL FUNGI CULTIVATION

Most people who have never attempted to grow mushrooms are blissfully unaware of how tedious and meticulous the process can be. The growth cycle isn't anything like growing a plant from a seed unless, of course, you're in the business of micro-propagation and GMOs. Even then, the only similarity is that they both use nutrient-rich media to support growth.

A makeshift flow hood I made for tissue culture

As stated earlier in the "Author's Note" at the beginning of the book, one of my many endeavors as a mycology hobbyist was growing fungi artificially from scratch (or spores). The initial success with growing oyster mushrooms using recyclable material of course was my motivation for starting new projects in mushroom cultivation. However, this time I wanted to get it right the first time, and I wanted to do it 'by the books', so to speak. So I read through several cultivation books such as *Mushroom Growers Handbook* and *Small-Scale Mushroom Cultivation*; both extremely helpful books that I would recommend to any first time growers. Then, it was on to the process from step one.

Step one can be carried out in one of two ways: 1) A method which utilizes Potato Dextrose Agar (PDA) media and a mushroom tissue sample, or 2) through spores.

Materials required for Step 1, method #1 are:

1) A laminar flow hood
2) A petri-dish (complete with PDA media)
3) A new or sterilized scalpel
4) An alcohol burning lamp (preferred over lighter as you don't need to touch it)
5) The mushroom you wish to cultivate
6) Protective covering to prevent contact contamination (a painter's mask, latex gloves, lab coat)
7) Rubbing alcohol to ensure both work area and tools are sterilized (overkill is underrated when it comes to work with PDA)

Fungi tissue culture in a petri dish
Photo by Alan Rockefeller

PDA is a nutrient mixture that works very well for the beginning stages of fungal growth; however, because of this, it supports all types of fungal growth even unintended airborne fungal spores. Due to this factor, when using PDA, the conditions that you work in must be aseptic; the work environment, as well as the person handling the materials must be sterilized and free of contamination. In order to do this, mycologists use what is called a "flow hood". A flow hood is essentially a box in which air is constantly being purified of contamination and being filtered throughout the workspace. This reduces the risks of airborne contaminates, such as mold spores, coming into contact with the media.

After hands (with gloves) are sterilized with rubbing alcohol, the petri-dish with the PDA media is placed under the flow hood (sealed) and the hood turned on. The scalpel is unpackaged (unless one has been previously sterilized) and then held over a flame until red-hot. The mushroom being used is then cut in half, and the knife held over the flame again to quickly re-sterilize; the flesh of the mushroom is then cut at a thick point (generally on the stem just before the cap) and the piece of flesh is dropped into the petri-dish; the petri-dish is then sealed with parafilm to allow gas exchange and prevent contaminates from entering while being stored. The finished product should be stored in a cool dark area, generally at about 20-23 °C (depending on the species), until the mycelium covers the media completely. When the media is covered by mycelium, it is ready to go on to the next step – grain spawn inoculation. Note that this method is usually used by advanced mushroom growers that will be sustaining a moderate to large grow operation, and is generally inadvisable for first time growers due to expenses of equipment and materials.

Materials required for Step 1, method #2 are:

1) An unpackaged syringe
2) A spore print
3) Boiled water
4) A sterilized knife

When using spores, the first thing that you would want to do is make a spore syringe. The purpose of utilizing a syringe as the vessel of spore delivery is that the small surface area of the needle will reduce the risk of contamination when inoculating the spawn jar. In order to make a spore syringe, one must first obtain a spore print (as explained on pg. 132). When the spore print is ready, water must be boiled for approximately 15 minutes to sterilize any contaminants and then when cooled down, poured into a jar that has been sterilized with alcohol (small mason jars work fine). The spores from the print are then scraped into the water with a sterilized knife or tool, and then stirred. It is best to make several spore prints when making a syringe; the more spores, the better. After the water has been stirred, insert a newly unpackaged syringe and suck up the spore-filled water. Now you're ready for inoculation!

Step Two will be done in two ways as well, regarding how you did the first step. Both use grain spawn in a sterilized jar, though only the method using PDA requires a flow hood.

Materials required for Step 2, method #1 (following the petri-dish method) are:

1) A laminar flow hood
2) Grain: wheat, rye, brown rice, white rice, wild bird seed or popcorn (out of these, rye grains are best choice)
3) Gypsum powder (optional but helpful)
4) Mason Jars
5) A pressure cooker (large enough to fit mason jars inside and close the lid)
6) A new or sterilized scalpel
7) Tin foil
8) The petri-dish with mycelium from step

Grain spawn jars are the seeds of the mushroom grow operation, and to prepare these jars, as in the first steps, the conditions must be contamination-free. Although rye grains are the standard for making grain spawn jars, other grains (as listed in materials) will work as well (depending on the species being grown).

In order to make these spawn jars, they grain and jars must be sterilized. However, before sterilization the grains should be soaked rinsed a few times to get rid of dirt and other unwanted materials and then soaked for about 24 hours. Once finished soaking, the grains should be placed in a pot, brought to boil and then allowed to sit and simmer in the pot for 15 minutes (this will kill off any unwanted sprouts). When simmering is finished, spread the grains out on foil to dry for an hour or two (note: some people like to sprinkle and mix a tablespoon of gypsum with their grains so that they have less chance of sticking to each other, this method can be very helpful but not entirely necessary); they should now be ready for the sterilization process!

To sterilize the grains, they should be loaded into a mason jar; about 60% of the jar should be filled. The tops of the jars are then covered with tinfoil placed into a pressure cooker at 15 PSI for 90 minutes. If there is no shelf in the pressure cookers, it's best to put mason jar lid frames from mason jars down so the glass doesn't contact the metal bottom directly, also, its best to keep the lid slightly undone to prevent too much pressure build-up in the jars. When the jars are removed, they are allowed to cool down for some time, before they are ready to be inoculated. Inoculation of the jars is done with the mycelium covered PDA in the petri-dish from step one. While wearing protective clothing and sterilized gloves, the flow hood is turned on and the spawn jar and petri-dish are opened underneath it. A slice of mycelium-covered PDA is taken using a sterilized scalpel and placed into the grain jar. Both the jar and petri-dish are then closed to prevent any further risk of airborne contamination. The spawn jar is then shaken to allow the PDA to be encompassed by the grains.

The spawn jar should be put in a cool, dark area at about 28-30°C, to allow the mycelium to spread over and engulf the grains; this part of the process is called the spawn run. The jars should be shaken every few days after mycelium starts to spread in order to keep grains from sticking to each other, as they need to remain "seed-like" for the next step of the process.

Healthy grain spawn created by inoculating rye grains in a glass bottle. Sometime the bottle is covered in tape to prevent light entry

Materials required for Step 2, method #2 (following the spore syringe method) are:

1) Vermiculite
2) Brown rice flour
3) Pressure cooker
4) Mason jars
5) Coffee filters
6) Tin foil
7) Rubber bands
8) The spore syringe from step 1
9) A plastic container
10) Perlite

This process bypasses the grain spawn process and goes directly into the grow phase by making rice-vermiculite "cakes". This type of cultivation is meant for small-scale and first time grows because it is much less tedious than the suspension methods and doesn't require expensive equipment and materials.

In order to make these cakes, 1 part vermiculite is added to 1/2 parts brown rice flour and 1/2 parts water, and then mixed. The mixture is put into a mason jar (about 90% of the jar should be filled), 2-3 small holes are poked in the lid, the top is covered with tin foil and then put into a pressure cooker to be sterilized at 15 PSI for 90 minutes. Again, if there is no shelf in the pressure cookers, it's best to put mason jar lid frames down so the glass doesn't contact the metal bottom directly, and keep the lid slightly undone to prevent too much pressure build-up in the jars. When finished and cooled down, the spore syringe is injected into the substrate to inoculate with spores (1 - 2ml per jar is sufficient), the more surface area covered, the better; this is why multiple holes are made, rather than one. After inoculation, a coffee filter is place over top of the lid and secured with a rubber band in order to allow gas exchange but prevent any contamination through the holes. The jar should be left in a cool, dark area at about 28-30°C until the mycelium has visibly engulfed the substrate; when this happens, the cake should be extracted from the mason jar, either by cracking the glass, or, if the jar has no curvature at the lid, sliding it out; this process is generally referred to as "birthing". Once it is free from the glass container, it is placed in on the lid inside a plastic container with holes in the middle and bottom for gas exchange.

The container or "fruiting chamber" will be filled with Perlite that has been soaked in water, this will retain and distribute moisture to throughout the container. This container should be kept in a cool place at about 28-30°C with dim light or indirect sunlight and should be sustained at 90-95% humidity levels. After some days, the cake will start pinning and the fruiting bodies will begin to grow. Thus, concluding the process for method #2.

Step 3 is only necessary for method #1 in order to expand the mushroom growing operation.

Materials required for Step 3, method #1 are:

1) Plastic bags (clear or black)
2) Substrate (i.e. spent coffee grounds, wood shavings, straw)
3) Cotton wool or yarn
4) Rubber bands
5) Plastic rings or PVC pipe cut at 1inch
6) Grain spawn made in Step 2
7) An autoclave

Substrate bags inoculated with grain spawn

In order to make these bags, substrate must first be pasteurized. Similar to the other steps in sterilizing substrate, it must be heated at high temperatures for some time in order to kill off any pre-existing contaminants. Typically, pasteurization is done with an autoclave, though substrate can also be pasteurized in a large vat or over a fire in a metal container. When the pasteurization is complete, the substrate is put into bag; in between every few scoops of substrate a bit of healthy grain spawn is sprinkled within, as to cover more surface area for growth. The top of the bag is then pushed through the PVC ring and is stuffed with cotton wool to enable gas exchange. The top of the bag is secured with a rubber band to keep the ring and wool in place. The bags are kept in a cool, dark room and allowed to incubate until the mycelium has engulfed the substrate. When the mycelium has completely overrun the substrate, the bags are then moved to another cool, dark room with a humidifier or some form of constant humidity. Cool humid conditions promote fruiting body growth. This 3 step method is very beneficial because the single PDA petri-dish of mycelium can seed up to about 10 large spawn grain jars and those 10 jars can seed up to about 20 mushroom growing bags, each. This is what I mean when I say that this method is for large-scale mushroom operations. If you are serious about starting a commercial mushroom farm, this is the method that would I would recommend, as making hundreds of rice cakes is not a feasible choice for a large yield.

Fruiting substrate bags in a local mushroom farm

PLUGS & LOG CULTIVATION

Simplicity is usually favored when starting new projects. A more convenient and likely more feasible method of cultivation here in Belize may be through plugging logs. The term "plugging" refers to a method of growing mushrooms by inserting wooden dowels into a host log. The first steps of this process are similar to the inoculation of grain spawn; however, instead of grains, the mycelium engulfed substrate is a wooden dowel. The process of making these plugs is similar to the grain spawn process already explained, with small wooden dowels replacing the grains.

Once the plugs are ready, you must find a log to inoculate. This part can be tricky, as the species of mushroom has to be able to thrive in the type of wood that you've chosen. A bit of extra research may be required in order to choose the best log as substrate to grow. I would suggest going into the forest and finding the edible (or medicinal) species of mushroom that you wish to grow and try to document the type of tree that it is growing on. Once you have documented several instances of the species growing, find out if any of these woody substrates are reoccurring. Is it only one type of tree the mushrooms appear to favor over another, or does the mushroom occur on several different tree species?

When you have determined the proper log to use for inoculation, you must drill holes in the wood, deep enough to insert the mycelium plugs. Once the pugs have been inserted, melt some bees wax and seal the holes. Bees wax is the standard for log inoculation because it not only seals the holes; it can also be consumed by healthy mycelium when the fruiting bodies are ready to emerge.

Shiitake mushrooms growing on wood.
Growing shiitake mushrooms (*Lentinulaedodes*) is commonly done through log cultivation.

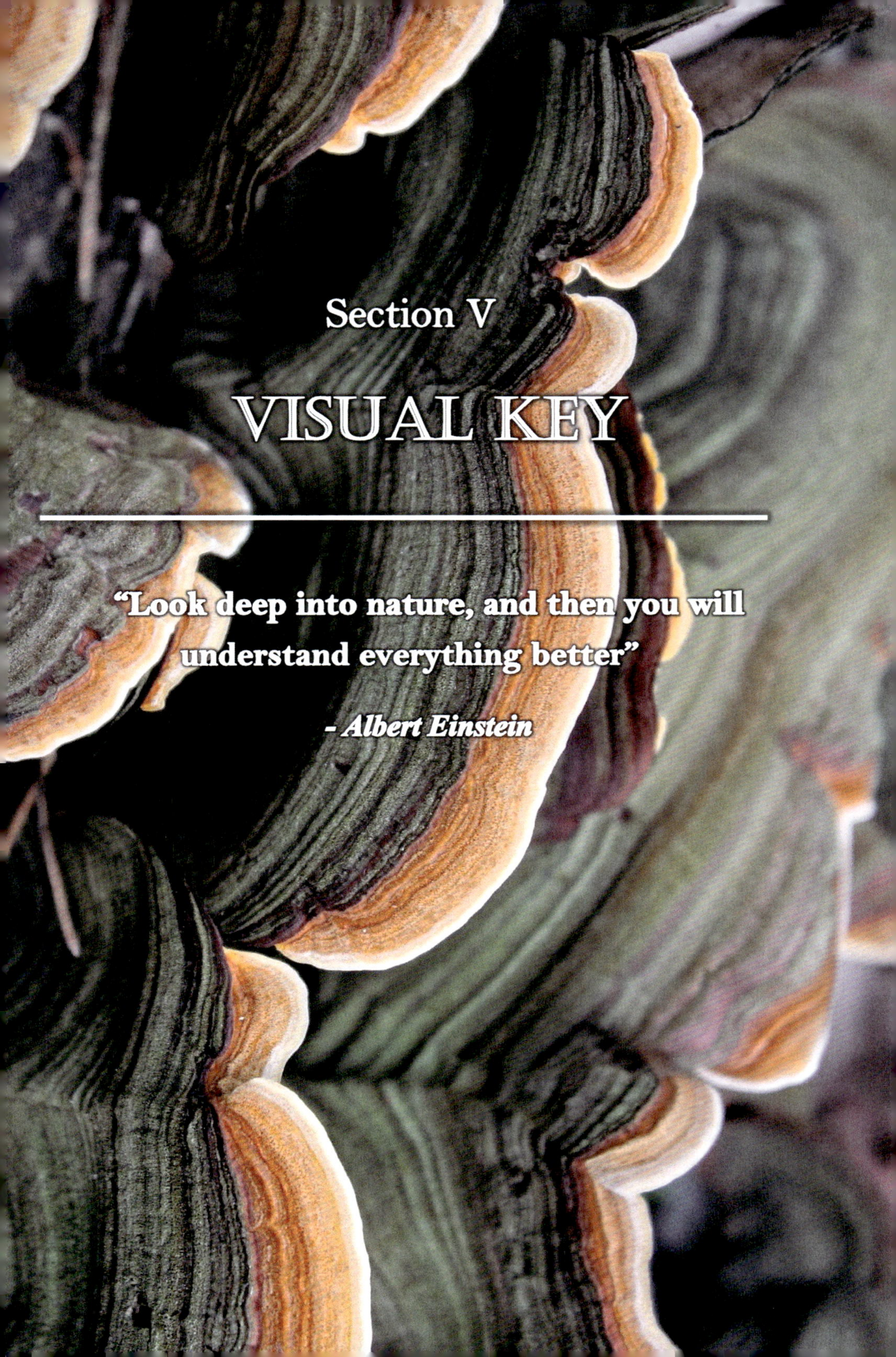

Section V

VISUAL KEY

"Look deep into nature, and then you will understand everything better"

- Albert Einstein

Golden Stalked Rosette (*Cotylidia aurantiaca*)

Chicken of the Woods (*Laetiporus caribeaus*)

Wine Glass Fungi (*Podoscypha petalodes*)

Stalked Polypore (*Amauroderma* sp.)

Tropical White Polypore (*Favolus tenuiculus*)

Layered Polypore (*Ganoderma australe*)

Artist Bracket (*Ganoderma applanatum*)

Reishi Mushroom (*Ganoderma* cf. *resinaceum*)

Varnish Shelf (*Ganoderma* sp.)

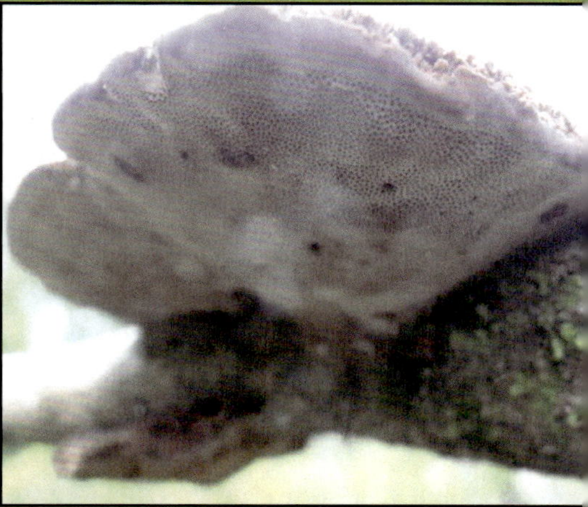

Hairy Hexagonia (*Hexagonia hydnoides*)

Weeping Polypore (*Inonotus dryadeus*)

Shaggy Bracket (*Inonotus hispidus*)

Dye-makers Fungus (*Phaeolus schweinitzii*)

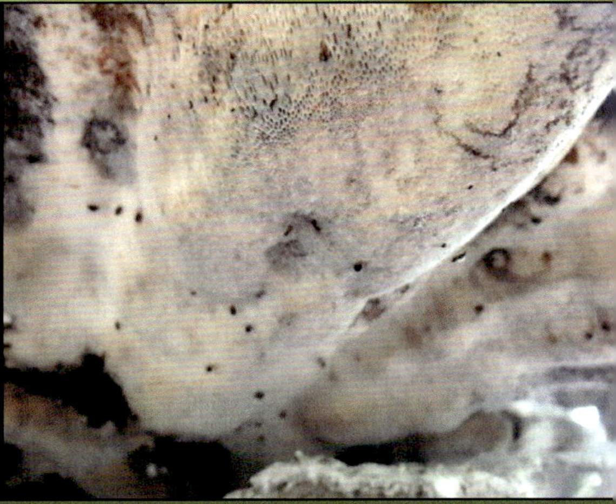

Big Brown (*Polyporus brittonii*)

Black-footed Polypore (*Polyporus tubaeformis*)

Blood-colored Polypore (*Pycnoporus sanguineus*)

Woodcrust Fungus (*Stereum* sp.)

Lumpy Bracket (*Trametes gibbosa*)

Hairy Bracket (*Trametes* cf. *hirsuta*)

White Lettuce Polypore (*Trametes membranacea*)

Turkey Tail (*Trametes versicolor*)

White Oysterling (*Claudopus depluens*)

Shoehorn Oyster (*Hohenbuehelia petaloides*)

Maze Gill (*Lenzite betulina*)

Oyster Mushroom (*Pleurotus* spp.)

Angel's Wings (*Pleurocybella porrigens*)

Split Gill (*Schizophyllum commune*)

Meadow Mushroom (*Agaricus* cf. *campestris*)

Woodland Mushroom (*Agaricus semotus*)

Common Fieldcap (*Agrocybe pediades*)

Jackson's Slender Caesar (*Amanita jacksonii*)

Fly Agaric (*Amanita muscaria* var. *flavivolvata*)

Ringless Honey Mushroom (*Armillaria tabescens*)

Netted Fieldcap (*Bolbitius reticulus*)

Indigo Velvetcap (*Calliderma* cf. *caeruleocapitata*)

Yellowfoot Chanterelle (*Cantharellus tubaeformis*)

Velvet Beetlecap (*Cantharocybe* sp.)

Pod Parachute (*Caripia/Gymnopus montagnei*)

The Vomiter (*Chlorophyllum molybdites*)

Fairy Inkcap (*Coprinellus disseminatus*)

Freckled Inkcap (*Coprinellus flocculosus*)

Shaggy Inkcap (*Coprinus comatus*)

Snowy Waxycap (*Cuphophyllus* cf. *virgineus*)

Thorncap Entoloma (*Entoloma dragonosporum*)

Rustgill Mushroom (*Gymnopilus parvulus*)

Oak-loving Gymnopus (*Gymnopus dryophilus*)

Slender-stalk (*Heimiomyces/Xeromphalina tenuipes*)

Stainless Witch's Hat (*Hygrocybe* cf. *noninquinans*)

Emerald Waxycap (*Hygrocybe* aff. *chloochlora*)

Crimson Waxycap (*Hygrocybe* cf. *occidentalis*)

Indigo Milkycap (*Lactarius indigo*)

Orange Milkycap (*Lactarius salmoneus*)

Hairy Funnelcap (*Lentinus crinitus*)

Stinking Dapperling (*Lepiota cristata*)

Plantpot Dapperling (*Leucocoprinus birnbaumii*)

White Dapperling (*Leucocoprinus cretaceus*)

Fragile Dapperling (*Leucocoprinus fragilissimus*)

Pink Pinwheel (*Marasmius haematocephalus*)

Tropical Pinwheel (*Marasmius plumieri*)

Orange Pinwheel (*Marasmius siccus*)

Sullivant's Pinwheel (*Marasmius* cf. *sullivantii*)

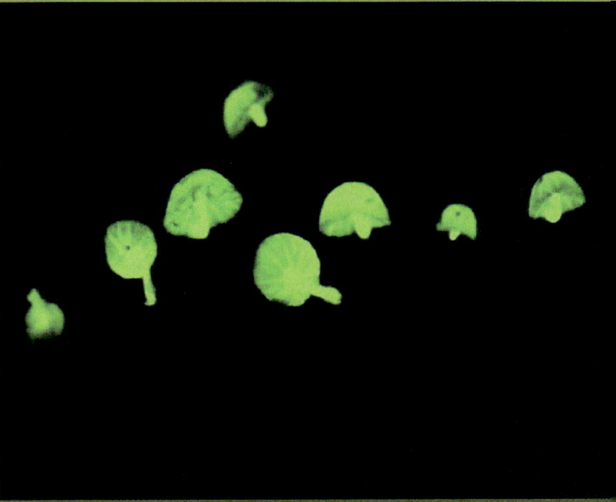

Glowing Bonnet (*Mycena margarita*)

The Train Wrecker (*Neolentinus lepideus*)

Common Mottlegill (*Panaeolus antillarum*)

Banded Mottlegill (*Panaeolus cinctulus*)

Blue Meany (*Panaeolus cyanescens*)

Petticoat Mottlegill (*Panaeolus papilionaceus*)

Shaggy Scalycap (*Pholiota squarrosa*)

Common Brittlestem (*Psathyrella candolleana*)

Hairy Brittlestem (*Psathyrella hirta*)

Magic Mushroom (*Psilocybe cubensis*)

Golden Brittlegill (*Russula flavida*)

Butterfoot Bolete (*Aureboletus/Boletus auriporus*)

Pineapple Bolete (*Boletellus ananas*)

Dupain's Bolete (*Boletus/Rubroboletus dupainii*)

Cracked-Cap Bolete (*Boletus variipes* var. *fagicola*)

Wrinkled Bolete (*Leccinum viscosum*)

Powdery Sulfer Bolete (*Pulveroboletus ravenelii*)

Latticed Stinkhorn (*Clathrus crispus*)

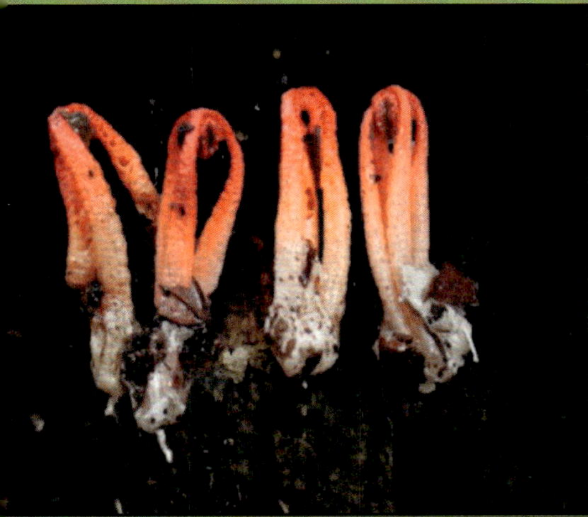
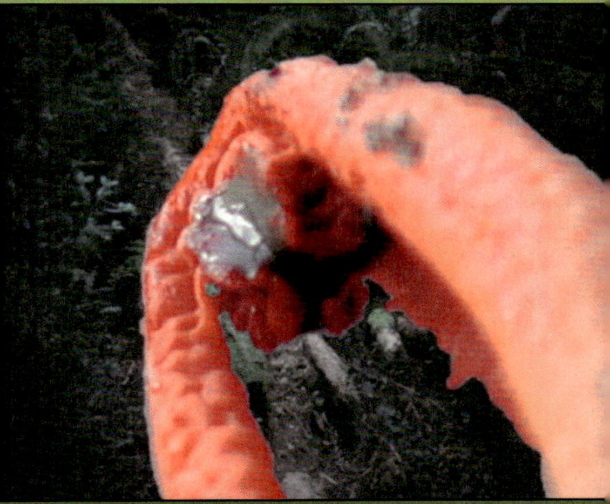

Lantern Stinkhorn (*Laternea triscapa*)

Veiled Stinkhorn (*Phallus indusiatus*)

Jelly Ear Fungus (*Auricularia auricula-judae*)

Tripe Fungus (*Auricularia mesenterica*)

Fan-shaped Jelly Fungus (*Dacryopinax spathularia*)

Snow Fungus (*Tremella fuciformis*)

Cloud Ear Fungus (*Auricularia nigricans*)

Pink Goblet (*Cookeina speciosa*)

White Eyelash Cup (*Cookeina tricholoma*)

Barnyard Cup (*Peziza vesiculosa*)

Tropical Bird's Nest Fungi (*Cyathus berkeleyanus*)

Cordycepioid Fungi (*Akanthomyces/Ophiocordyceps* sp.)

Blue-tipped Coral (*Phaeoclavulina cyanocephala*)

Thin-branch Coral (*Lentaria surculus*)

King Alfred's Cake (*Daldinia concentrica*)

Candlesnuff Fungus (*Xylaria hypoxylon*)

Dead Man's Finger (*Xylaria polymorpha*)

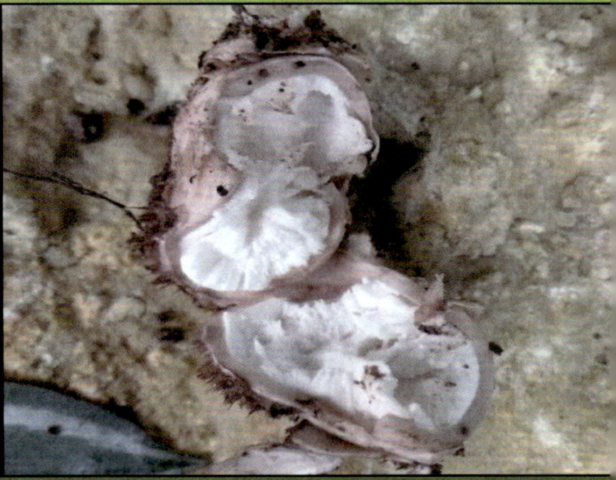

Spiny Puffball (*Lycoperdon echinatum*)

Common Puffball (*Lycoperdon perlatum*)

Stump Puffball (*Lycoperdon* cf. *pyriforme*)

Dead Man's Foot (*Pisolithus arhizus*)

Rounded Earthstar (*Gaestrum saccatum*)

Berkley's Earthstar (*Geastrum berkeleyi*)

Carnival Candy Slime Mold (*Arcyria denudate*)

Bell-shaped Slime Mold (*Physarella oblonga*)

Chocolate Tube Slime (*Stemonitis splendens*)

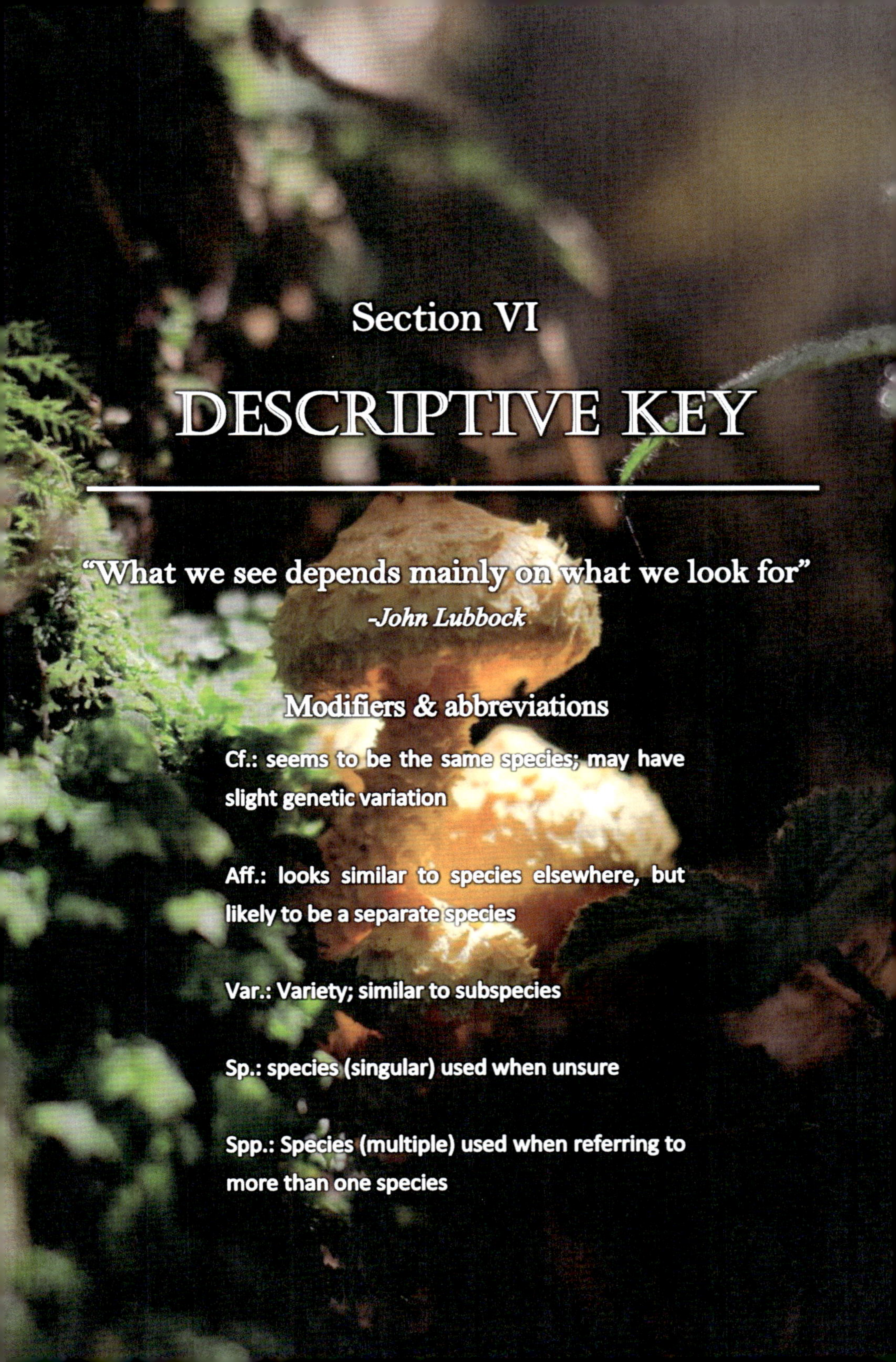

Section VI

DESCRIPTIVE KEY

"What we see depends mainly on what we look for"

-John Lubbock

Modifiers & abbreviations

Cf.: seems to be the same species; may have slight genetic variation

Aff.: looks similar to species elsewhere, but likely to be a separate species

Var.: Variety; similar to subspecies

Sp.: species (singular) used when unsure

Spp.: Species (multiple) used when referring to more than one species

Rosette Polypores

Golden Stalked Rosette *(Cotylidia aurantiaca)*

Description: A tiny fungus, usually growing in groups; fruiting bodies are paper thin, fan shaped, and yellow in color; margin is usually ridged or serrated

Spore Print: Unknown

Habitat & Preferred Substrate: Found mostly in broadleaf and dense canopy forests of Belize; saprotrophic; grows on organic matter, usually chunks of rotting wood.

Edibility: Inedible.

Chicken of the Woods *(Laetiporus caribeaus)*

Description: A relatively large, soft bracket fungus found in overlapping clumps. Upper side is yellow to orange (sometimes both), smooth and wavy; pore surface is yellow and features small pores; no stalk present.

Spore Print: White.

Habitat & Preferred Substrate: Found throughout Belize; however, not very common; saprophytic; grows on dead and rotting wood.

Edibility: Edible.

Similar Species: *Pycnoporus cinnabarinus* is also orange to reddish in color, though does not cluster and fuse as do *Laetiporus spp*.

Note: There are at least two known species of Chicken of the Woods fungi that occur in Belize: *L. gilbertsonii* and *L. caribensis* (Banik et al., 2012).

Wine Glass Fungi *(Podoscypha petalodes)*

Description: A thin bracket fungus with pink, orange, yellow or white fruiting bodies; young fruiting bodies look wine-glass shaped, as the common name suggests, and turn large and rosette-shaped when mature; pore surface and upper surface are the same color; stipe very short, if present.

Spore Print: White

Habitat & Preferred Substrate: Found mostly in broadleaf and dense canopy forests of Belize; saprophytic; grows on dead and rotting wood.

Edibility: Inedible.

Bracket Polypores

Stalked Polypore *(Amauroderma* sp.)

Description: A small to large sized polypore with a tough or woody texture; fruiting bodies have a round or kidney shaped, dry cap with color ranging from black, tan, reddish, and grey; central zone of cap is usually dark brown to black in color; stalk can be medium to long, thick or thin, depending on species

Spore Print: Unknown

Habitat & Preferred Substrate: Found in moist forests of Belize on the ground or on trees; saprotrophic; grows on decaying trees or organic matter in the soil.

Edibility: Inedible

Tropical White Polypore *(Favolus tenuiculus)*

Description: White fruiting body with glossy (unless dried out), kidney-shaped cap; stalk is short and usually converges with the substrate. The underside has net-like elongated pores which descend and end at the stalk; stalk is short and matches color of the fruiting body.

Spore Print: White

Habitat & Preferred Substrate: Usually in broadleaf forest areas; saprophyte; grows on dead and rotting wood.

Edibility: Edible

Layered Polypore (*Ganoderma australe*)

Description: A tough woody bracket fungus. Cap shape appears as if multiple brackets stacked on top of eachother; cap color is usually light to dark brown and powdery, margin usually has a white brim; pore surface is white to brown in color depending on maturity, usually flat, and features small pores; stalk is merged with substrate and otherwise non-existent; white pore surface bruises brown.

Spore Print: Brown

Habitat & Preferred Substrate: Parasitic/Saprophytic; grows on the sides of living and dead trees in moist forests of Belize.

Edibility: Inedible.

Artist Bracket (*Ganoderma applanatum*)

Description: A tough woody bracket fungus; cap is semi-circular to irregular; smooth, lumpy, and grooved; cap surface is white when immature, but soon turns a light cinnamon-brown color; inner flesh is light brown; Stalk is merged with substrate and otherwise non-existent. The name "Artist Bracket" comes from the white pore surface that stains brown when scraped or pressed. Some have been used to make intricate sketches that last for years.

Spore Print: Brown

Habitat & Preferred Substrate: Found commonly throughout Belize; saprophytic/parasitic; grows on dead and living hardwoods.

Edibility: Inedible raw; has been said to be used in tea and tinctures for potential medicinal purposes.

Reishi Mushroom (*Ganoderma* cf. *resinaceum*)

Description: A tough woody fungus with a tough woody stalk (if present); Cap is varnished (shiny) with descending, lumpy zones; color is usually reddish-purple and gets lighter towards the margin until it turns white; pore surface is white, and becomes brown with age; inner flesh is brown; stalk is tough, reddish to purplish and varnished in appearance.

Spore Print: Brown.

Habitat & Preferred Substrate: Found throughout Belize; saprophytic/parasitic; grows on dead and living hardwoods.

Edibility: Medicinal; can consume in tea and tincture; research is being carried out on immune assisting and anti-cancer properties.

Varnish Shelf (*Ganoderma* sp.)

Description: A tough woody fungus with a tough woody stalk (if present); Cap is varnished (shiny) with descending, lumpy zones; color is purple even at immature stages; pore surface is flat and dark brown, and becomes brown with age; inner flesh is brown; stalk is tough, reddish to purplish and varnished in appearance.

Spore Print: Brown

Habitat & Preferred Substrate: Found in hardwood forests; saprophytic/parasitic; grows on dead and living hardwoods.

Edibility: Medicinal; can consume in tea and tincture

Note: This specimen is believed to be Ganoderma fornicatum (sensu lato). Unfortunately, many Ganoderma species look identical and cannot be told apart without microscopy.

Hairy Hexagonia (*Hexagonia hydnoides*)
Description: A tough woody bracket fungus with a fuzzy dark brown to black cap surface and a dark brown pore surface, featuring small pores; stalk is merged with substrate and is otherwise non-existent.
Spore Print: White.
Habitat & Preferred Substrate: found in both open and closed canopy areas of Belize; saprophytic/parasitic; grows on dead and some live trees.
Edibility: Inedible.

Weeping Polypore *(Inonotus dryadeus*)
Description: A slightly tough, thick bracket fungus; cap is irregularly shaped, lumpy, velvety in texture, and dingy yellow to brownish in color; surface of fruiting body secretes a reddish brown liquid; pore surface is dingy yellow to brownish in color and also secretes reddish brown liquid; fruiting body is attached to substrate; stalk is not present.
Spore Print: Yellow to brown
Habitat & Preferred Substrate: Found in shaded forest areas; parasitic; grows on the buttresses of live and some dying or rotting trees in Belize.
Edibility: Inedible

Shaggy Bracket *(Inonotus hispidus*)
Description: A tough woody bracket fungus with an orange to light brown fuzzy cap surface and a white to yellow margin and pore surface featuring small pores; stalk is merged with substrate and is otherwise non-existent.
Spore Print: Brown.
Habitat & Preferred Substrate: found in open or partly shaded forest areas; parasitic; found on ash trees (*Fraxinus dubia).*
Edibility: Inedible

Dye-makers fungus (*Phaeolus schweinitzii*)
Description: A tough woody bracket fungus with a black lumpy cap surface; off-white margin leads to a white to yellow/greenish yellow pore surface; Stalk is merged with substrate and is otherwise non-existent. As the name suggests, this fungus is used in the production of colored dyes.
Spore Print: White to yellow
Habitat & Preferred Substrate: Found in forested and open areas of Belize; parasitic; grows on the sides of trees.
Edibility: Inedible or poisonous.

Big Brown *(Polyporus brittonii)*

Description: A very large, thick polypore fungus; cap surface is velvety, lumpy, broadly rosette, and purplish brown in color with some darker and lighter bands proximal to the margin (cap color can be variable when covered in moss or weathered); pore surface is white; stalk is merged with substrate and is otherwise non-existent.

Spore Print: White

Habitat & Preferred Substrate: Found in moist forests of Belize; saprophytic; grows on dead and rotting wood.

Edibility: Edible

Black-footed Polypore (*Polyporus tubaeformis*)

Description: A tough, thin bracket fungus; cap is funnel shaped, and orange, light brown to greyish in color, occasionally with a darkened center; pore surface is white to light reddish brown when mature or weathered; stalk is short and black, as the common name suggests.

Spore Print: White

Habitat & Preferred Substrate: Found in moist forests of Belize; saprotrophic; grows on dead and rotting wood.

Edibility: Inedible

Blood colored polypore (*Pycnoporus sanguineus*)

Description: Cap flat or wavy; smooth/velvety depending on age; kidney-shaped; orange to red in color, sometimes with a white or yellow brim at the margin; pore surface features red to orange pores; stalk is merged with substrate and is otherwise non-existent.

Spore Print: white.

Habitat & Preferred Substrate: Common throughout Central and Southern Belize; saprophytic; grows on dead and rotting wood.

Edibility: Inedible

Woodcrust Fungus *(Stereum* sp.)

Description: A clustering, fused woodcrust fungus; fruiting body is irregularly shaped, sometimes wavy, and often features overlapping projections; upper surface is white, light orange to reddish, sometimes with lighter banded zones proximal to the margins; pore surface is absent of visible pores, smooth, and orange, pinkish, to brownish red in color; stalk not present; texture is rubbery, gelatinous or tough depending on maturity.

Spore Print: White
Habitat & Preferred Substrate: Found in moist, closed canopy forest areas of Belize; saprophytic; grows on rotting hardwoods.
Edibility: Inedible

Lumpy Bracket (Trametes gibbosa)

Description: A tough, woody bracket fungus; cap is semicircular with multiple lumpy zones; upper surface is white, to off-white, often turning green with algae when mature; pores on pore surface are elongated, and white to light yellow in color
Spore Print: White
Habitat & Preferred Substrate: Found in moist forests of Belize; saprophytic; grows on dead and rotting wood.
Edibility: Inedible

Hairy Bracket (*Trametes* cf. *hirsuta*)

Description: A tough, woody bracket fungus; fruiting body is semi-circular with lumpy zones and fine, velvety hairs on the upper surface; cap surface is white to brownish, occasionally featuring various colored zone bands (can appear green when covered in algae); pore surface is white, turning yellow to brownish when mature; stalk is merged with substrate and is otherwise non-existent.
Spore Print: White
Habitat & Preferred Substrate: Found in moist, closed canopy forests of Belize; saprophytic; grows on dead and rotting hardwoods.
Edibility: Inedible

White Lettuce Polypore *(Trametes membranacea)*

Description: A thin, chip-like, clustering bracket fungus; fruiting bodies are fan shaped, white to yellowish in color, and have ridged or serrated margins, pore surface is white; stalk is merged with substrate and is otherwise non-existent; found almost exclusively overlapping aggregations.
Spore Print: White
Habitat & Preferred Substrate: Found in moist forests of Belize; saprophytic; grows on dead and rotting wood.
Edibility: Inedible

Turkey Tail *(Trametes versicolor*)

Description: A tough bracket fungus which has many color variations and growth patterns; found in both clustered, rosette, and individual semi-circular fruiting bodies. Cap is smooth and flat or slightly depressed. As the name suggests, the upper side of the cap exhibits color variations that resemble that of a turkey's tail. Typical coloration of the changing rings or "zones" on the upper side of the cap are black, yellow and rust brown; however, some can be covered in green algae, and some can be devoid of color due to weathering. The pores underside appears white, yellow or light brown; Stalk is merged with substrate and is otherwise non-existent. Fruiting bodies are thin.

Spore Print: White.

Habitat & Preferred Substrate: Common throughout Belize; saprophytic; grows on dead and rotting wood.

Edibility: Although this fungus is not generally considered edible, it is being studied for medicinal properties and can be used in tea and tincture.

Gilled Brackets

White Oysterling (*Claudopus/Entoloma depluens*)

Description: A small fungi usually growing in groups or clusters; cap is kidney or oyster shaped, white, and has thin fuzz on the upper surface; gills are close, adnexed, and, white, pink, to light brown; stalk is very short and white to light yellow; annulus not present.
Spore Print: Pinkish brown
Habitat & Preferred Substrate: Found in moist broadleaf forests of Belize; saprophytic; grows on dead and rotting hardwoods.
Edibility: Unknown

Shoehorn Osyter (*Hohenbuehelia petaloides*)

Description: A small oyster mushroom; fruiting body is thin, white and tan in color, and fan-shaped or oyster-like; stalk converges with substrate; underside features close to crowded ridges which descend to the base of the fruiting body where it attaches to the substrate; gill color is tan closest to substrate, becoming white to off-white proximal to the margin.
Spore Print: White
Habitat & Preferred Substrate: Commonly found in broadleaf forest areas; saprophytic; grows on dead and rotting wood.
Edibility: Edible

Maze Gill (*Lenzite betulina*)

Description: A tough leathery/woody bracket fungus; cap is flat or descending and zoned; color white to brown (although green when covered in algae); gills white, yellow or brown in color and seemingly "maze-shaped", as the common name suggests; Stalk is merged with substrate and is otherwise non-existent.
Spore Print: White.
Habitat & Preferred Substrate: Found in moist forests with heavy canopy; saprophytic; grows on dead and rotting wood.
Edibility: Inedible.

Oyster Mushroom (*Pleurotus* spp.)

Description: A small to relatively large gilled bracket fungus, usually found in clusters; fruiting body can be fan shaped, kidney shaped, and oyster shaped; most oyster mushroom species color ranges from white, to tan, to

greyish with the exception of *Pleurotus djamor* (right) which is pink. Gills or ridges are close to crowded, matching the color of the fruiting body, and run decurrent to the base of a short stalk; fruiting bodies are usually attached to substrate, sometimes with a false stalk.
Spore Print: White
Habitat & Preferred Substrate: Commonly found in broadleaf forest areas; saprophytic; grows on dead and rotting hardwood.
Similar Species: Angel Wings (*Pleurocybella porrigens*) grow primarily on conifers and look nearly identical to true oyster mushrooms. Other non-oyster lookalikes may have a true stalk.
Note: There are two species shown in these photos: Pleurotus djamor (pink oyster, left) and Pleurotus (cf.) ostreatus (white oyster, right).

Angel's Wings *(Pleurocybella porrigens)*
Description: White fruiting body with a thin, fan-shaped or oyster-like cap; stalk is short and usually converges with the substrate; the underside features close to crowded ridges which descend to the base of the stalk.
Spore Print: White
Habitat & Preferred Substrate: Commonly found in coniferous forest areas; saprophytic; grows on dead and rotting wood, mostly conifers.
Edibility: Once considered edible, though poisoning mortalities have occurred in people with renal dysfunction. Therefore, this fungus should now be treated as poisonous.
Similar Species: Other edible oyster mushroom species

Split Gill (*Schizophyllum commune*)
Description: A small gilled bracket fungus with tough or rubbery texture; cap surface is velvety to finely hairy, irregularly fan-shaped, and white, tan, or greyish in color; gills are close, distinct, and split in the direction of each fan projection; gill color white to grey; Stalk is merged with substrate and is otherwise non-existent.
Spore Print: White
Habitat & Preferred Substrate: Found commonly throughout Belize; saprophytic; grows on dead and rotting wood
Edibility: Used by some ethnic groups as a sort of chewing gum and occasionally as food, though this species is generally not considered palatable.

Gilled Mushrooms

Meadow Mushroom (*Agaricus* cf. *campestris*)

Description: A medium sized, white to pale brown mushroom; cap is convex and flattens when mature; gills are crowded and pink when young, turning reddish brown to dark brown when mature; Stalk is medium sized and white
Spore Print: Reddish-brown to dark brown
Habitat & Preferred Substrate: Found commonly in human disturbed areas of Belize; saprotrophic; grows on organic matter.
Edibility: Edible
Similar Species: BEWARE - Poisonous *Agaricus sp.* with pink or brown gills can look almost identical to *A. campestris*. However, there is a simple way to determine whether this is the edible pink gilled meadow mushroom or a poisonous lookalike, and that is to check for yellow staining properties. When a poisonous lookalike is cut in half, the flesh within the cap, stem, or both, will stain yellow. If there is no yellow staining, it is likely that you have the edible meadow mushroom.

Woodland Mushroom (*Agaricus semotus*)

Description: A medium sized, white to pale brown mushroom; cap is convex and flattens when mature; gills are crowded and pink when young, turning reddish brown to dark brown when mature; Stipe is medium sized and white
Spore Print: Reddish-brown to dark brown
Habitat & Preferred Substrate: Found in both wooded and human disturbed areas of Belize; saprotrophic; grows on organic matter.
Edibility: Edible
Similar Species: BEWARE – Poisonous lookalike *Agaricus xanthodermis* also stains yellow. Although I'm unsure if it has been documented in Belize, it's best to be safe and avoid consumption of this mushroom.

Common Fieldcap (*Agrocybe pediades*)

Description: A tiny mushroom; cap is small, convex, and white to light yellowish brown; veil remnant sometimes found on the margin of young fruiting bodies; gills are close, adnate to stalk, and off-white at first becoming brown when mature; stalk is short, thin and matches the color of the cap; annulus not present.
Spore Print: Brown

Habitat & Preferred Substrate: Found in human disturbed areas of Belize; Coprophilous/saprotrophic; grows on animal dung and other organic matter.
Edibility: Edible, but easily confused with other inedible or poisonous little brown mushrooms. For this reason, it is best to avoid consumption.

Jackson's Slender Caesar (*Amanita jacksonii*)

Description: A medium to large sized mushroom; cap is convex with a central umbo, shiny, and reddish orange in the center, becoming lighter proximal to the margin; margin has close striations; gills are crowded, free from stalk or adnexed, and yellow; stalk is relatively long, thinner proximal to cap, and yellow in color; annulus is present; young fruiting bodies emerge with an egg-shaped universal veil and the red cap protruding from the top.
Spore Print: White
Habitat & Preferred Substrate: Found in both deciduous and coniferous forests; mycorrhizal; a symbiotic organism with trees.
Edibility: Edible

Fly Agaric (*Amanita muscaria* var. *flavivolvata*)

Description: A medium to large sized mushroom; cap is convex to broadly convex when mature, shiny, and red with white to yellowish patches or 'freckles'; gills are crowded, free from stalk or adnexed, and white; stalk is relatively long, white in color, and thick, exhibiting concentric zones on the swollen base; annulus is present.
Spore Print: White
Habitat & Preferred Substrate: Found in both deciduous and coniferous forests; mycorrhizal; a symbiotic organism with trees.
Edibility: Poisonous/Psychoactive

Ringless Honey Mushroom *(Armillaria tabescens*)

Description: A small to medium mushroom often found in clusters; cap is smooth, convex and flattens out with age, sometimes with a visible umbonate center; cap color is typically a pale brown or "honey" color, as the common name suggests, and occasionally exhibits dark colored, patchy scales concentrated toward the center; gills are close, decurrent to stalk, and white to pinkish or off-white with age; stalk is white or brownish becoming lighter proximal to the cap; stalk length varies and is usually fuses to stalks of other fruiting bodies in the cluster; annulus not present.
Spore Print: White

Habitat & Preferred Substrate: Typically found in broadleaf forests of Belize. Parasitic; found on the buttresses of live trees.
Edibility: Although this mushroom is considered edible, it is not easily identifiable by appearance alone, and some of the lookalikes can be severely poisonous, some even deadly. For this reason, it is best to avoid consuming this mushroom.

Netted Fieldcap *(Bolbitius reticulatus)*

Description: A small mushroom; cap is smooth, occasionally slimy, and may feature a reticulated or vein-like pattern; cap color can be grey, pale brown, or purplish with a darkened center; cap shape is bell-shaped, to convex, to flat as it matures, featuring a small central umbo; striations originating on the margin can run part way or all the way to the central umbo; gills are close, free from stalk, and white, pink, light brown, to rusty as it matures; stalk is medium sized, relatively thin, whitish to pale brown in color, powdery, and features fine scales; annulus not present.
Spore Print: Rusty brown
Habitat & Preferred Substrate: Found in moist forest areas or even plant pots in Belize; saprotrophic; grows on organic matter.
Edibility: Poisonous
Similar species: This mushroom can look like several different species of *Pluteus* and *Gymnopus*, of which some may be edible. For this reason, it is best for beginning foragers to avoid consumption of all three genera in order to avoid poisoning.
Note: *Bolbitius aleuriatus* is now considered the same species as *Bolbitius reticulatus*. The primary difference between the morphology of the two is that *B. aleuriatus* is usually grey in color, rather than brown, and lacks the reticulated cap pattern seen in *B. reticulatus* (Kuo, 2007).

Indigo Velvetcap (*Calliderma* cf. *caeruleocapitata*)

Description: A relatively small mushroom; cap is convex, velvety and indigo blue in color; gills are close to crowded, decurrent to stalk, and white to light pinkish brown with maturity
Spore Print: Light brown
Habitat & Preferred Substrate: Found in Guanacaste National Park, Belize; Ecological role is unknown; suspected to be saprotrophic, as it was growing chunks of decaying wood.
Edibility: Unknown

Yellowfoot Chanterelle (*Cantharellus tubaeformis*)

Description: A relatively small mushrooms; cap is flat and umbilicate, dry to waxy, and dark yellow, orange to brownish in color; margin is wavy when mature; gills or ridges are distant, veiny, decurrent to stalk, and grey to brown in color; stalk is medium length and thickness, and orange to brown in color; annulus not present.

Spore Print: White

Habitat & Preferred Substrate: Found in coniferous forests of Belize; mycorrhizal; a symbiotic organism with coniferous trees.

Edibility: Edible

Velvet Beetlecap (*Cantharocybe* sp.)

Description: A relatively large mushroom; cap is velvety, black, and convex at first, becoming depressed when mature. Margin is rolled inward; gills are crowded, decurrent to stalk, and white to light brown; stalk is thick, becoming thin proximal to cap, scaly, and grey; annulus not present.

Spore Print: Not taken

Habitat & Preferred Substrate: Found at Toucan Ridge Ecology and Education Society (T.R.E.E.S), Stann Creek, Belize. The ecological role of this mushroom is unknown, though it is suspected to be saprotrophic.

Edibility: Unknown

Note: This mushroom is suspected to be a close relative or variety of *Cantharocybe brunneovelutina* described in "A new *Cantharocybe* from Belize with notes on the type of *Cantharocybe gruberi*" (Ovrebo, et al. 2011). Another suspected species is Lactarius aff. Lignyotus, due to the color, velvety cap, and gills being a close match; however the stalk lacks scales.

Pod Parachute (*Caripia/Gymnopus montagnei*)

Description: A small mushroom usually found in groups; cap is off-white to brown, goblet to trumpet shaped, with a flat to concave, pale colored top; gills not present; stalk is thin, and off-white to light brown.

Spore Print: Not taken

Habitat & Preferred Substrate: Found in moist forest areas of Belize; saprotrophic; grows on organic matter, usually chunks of decaying wood.

Edibility: Inedible

The Vomiter *(Chlorophyllum molybdites)*

Description: A relatively large, terrestrial parasol mushroom; cap large, flat, and white, with scattered brown scales concentrating toward the center; gills are crowded, free from stalk, and white to dark green with maturity; stalk is long, thin , and white with occasional scales near the base; annulus is present.

Spore Print: Green

Habitat & Preferred Substrate: Commonly found in lawns and open fields after rain; saprotrophic; grows on organic matter.

Edibility: Poisonous

Similar Species: The true parasol mushroom (*Macrolepiota procera*) that appears almost identical to this species is edible. The main distinguishing factor being that the poisonous species has a green spore print, whereas the edible species spore print is white.

Fairy Inkcap (*Coprinellus disseminates*)

Description: A tiny bell-shaped mushroom found in large groups; cap is white, convex, with close striations originating from the small discolored umbo in the center down to the margin; gills are close, adnate to stalk, and brown in color; stipe is thin and appears translucent; annulus not present.

Spore Print: Black

Habitat & Preferred Substrate: Found commonly in moist forests of Belize; saprophytic; grows on dead and rotting wood.

Edibility: Edible

Freckled Inkcap (*Coprinellus flocculosus*)

Description: A tiny mushroom; cap is light brown, bell-shaped with small white scaly "freckles"; when mature, cap turns black, oily, and flared; gills are crowded, free from stalk, and cream to grey to black in color as the fruiting body matures; stalk is short, thick and swollen toward the base; annulus not present.

Spore Print: Black

Habitat & Preferred Substrate: Found in moist areas of Belize. Commonly seen on soil within the mossy stones of Maya ruins; saprotrophic; grows on organic matter.

Edibility: Unknown

Shaggy Inkcap (*Coprinus comatus*)

Description: A medium to large sized mushroom; cap is white, cylindrical to bell shaped when mature, and covered with shaggy scales; cap becomes black and 'inky' with maturity; gills are crowded, free from stalk, and white to black with maturity; stalk can be short or long, smooth, white, and swollen toward the base.

Spore Print: Black

Habitat & Preferred Substrate: Found in human disturbed areas of Belize; saprotrophic; grows on organic matter, usually in soil.

Edibility: Edible

Snowy waxycap *(Cuphophyllus* cf. *virgineus)*

Description: A small to medium sized mushroom with a slimy or waxy appearance and texture; cap is convex, turning depressed as it ages; cap may appear semi-translucent and features distant striations near the margin; gills are ridged (fixed to cap and stipe), descending, distant and match the color of the cap; stalk is white, smooth, medium sized and tapers off toward the base; annulus not present.

Spore Print: White

Habitat & Preferred Substrate: Can be found in tall grass lawns and pastures in Belize and is said to be associated with moss. The ecological role of this species is unknown.

Note: This is species complex which is thought to contain several species.

Edibility: Inedible

Thorncap Entoloma *(Entoloma dragonosporum)*

Description: A tall-standing mushroom; cap is conical and brown to greyish, with a light rimmed margin; at the center of the cap is a protruding, thorn-like papillus structure; distant striations originating at the papillus run down to the margin; gills are distant, adnate or sub-adnexed to stalk, and grey to pale brown in color; stalk is thin, lengthy and greyish to pale brown in color; annulus not present.

Spore Print: Salmon pink

Habitat & Preferred Substrate: Found in moist forest areas with ubiquitous leaf-litter; saprotrophic; grows on organic matter.

Edibility: Unknown

Rustgill Mushroom *(Gymnopilus parvulus*)

Description: A small to medium sized clustering mushroom; cap is convex, pale brown, pale yellow to brownish-orange in color with a dry darker colored scaly surface; gills are close, adnate to stalk, and yellow to orange in color; stalk is medium sized, pale yellow, sometimes fibrous in appearance and bruises rusty brown; remnants of the veil may be seen on the margin of young fruiting body caps; annulus is small and sometimes not present.

Spore Print: Rusty orange to rusty brown

Habitat & Preferred Substrate: Found in areas of Belize with coniferous trees; saprophytic; grows on dead coniferous wood.

Edibility: Inedible. Other *Gymnopilus* species, such as *Gymnopilus spectabilis* are psychoactive, and some lookalikes such as *Galerina* spp. are deadly poisonous. For this reason, consumption of these mushrooms should be avoided.

Oak-loving Gymnopus (*Gymnopus dryophilus*)

Description: A common medium sized mushroom; Cap is convex, smooth and russet to light brownish-yellow; gills are close, adnexed or free and white to off-white in color; stalk is relatively long, thin and matches the cap color toward the base, becoming lighter proximally to the cap; annulus not present.

Spore Print: White to cream

Habitat & Preferred Substrate: This mushroom can be found almost anywhere in forested areas and even plant pots; saprotrophic; grows on organic matter.

Edibility: Edible

Slender-stalk *(Heimiomyce/Xeromphalinas tenuipes*)

Description: A fairly small mushroom; cap is, somewhat velvety, convex or flat, brownish orange with a darker color concentrating toward the center; margin has close striations; gills are close, adnexed to stalk and white to pale yellow in color; stalk is medium sized, somewhat thick, brownish orange with fine velvety hairs; annulus not present.

Spore Print: White

Habitat & Preferred Substrate: Found in moist hardwood forests of Belize; saprophytic; grows on dead and rotting hardwoods.

Edibility: Inedible

Stainless Witch's Hat *(Hygrocybe* cf. *noninquinans*)

Description: A small waxy or slimy mushroom; cap is conical with a central umbo, reddish orange to yellowish orange, becoming lighter proximal to the margin; margin has distant striations; gills are distant to close, free from stalk, and white to pale yellow; stalk is medium sized, fibrous, pale yellow and white at the base; annulus not present.
Spore Print: White
Habitat & Preferred Substrate: Found in moist, tall grass areas of Belize. The ecological role of this species is unknown.
Edibility: Unknown

Emerald Waxycap *(Hygrocybe* aff. *chloochlora*)

Description: A small waxy or slimy mushroom; cap is pale green to dark green, and depressed with a central hole; margin is white to pale yellow in color and striated; gills are semi-distant, adnate and descending to stalk, green with white to off-white edges; stalk is thick, hollow, medium length and green turning pale green or yellow toward the base.
Spore Print: White
Habitat & Preferred Substrate: Found in closed canopy moist forests of Belize; associated with moss covered rotting logs.
Edibility: Unknown

Crimson Waxycap (*Hygrocybe* cf. *occidentalis*)

Description: A small to medium sized mushroom with a waxy texture; cap is red with a central hole; gills are distant, adnexed to stalk, and pale yellow to red in color; stalk is short to medium in length, hollow and reddish in color, sometimes becoming yellow proximal to the cap; annulus not present.
Spore Print: White
Habitat & Preferred Substrate: Found in moist forest areas of Belize. Suspected to be mycorrhizal; a symbiotic organism with vascular plants.
Edibility: Edible
Similar Species: There are a couple other red *Hygrocybe* species in Belize, including *H. occidentalis* and *H.* cf. *miniatus*. *H. occidentalis* typically has a yellow stalk and gills; *H.* cf. *miniata* has a red stalk and red cap with no highlighted colors on the margin or stalk.

Indigo Milkycap (*Lactarius Indigo*)

Description: A medium sized mushroom; cap is convex with a slight depression, indigo blue to greyish blue in color, and exhibits concentric zones; gills are close, adnate to stalk or slightly decurrent and indigo blue in color with pale edges; gills secrete blue latex which becomes greenish when dry; stalk is short, thick and indigo blue; inner flesh stains blue when injured.

Spore Print: Cream

Habitat & Preferred Substrate: Found mostly in coniferous forests and mixed deciduous ecotone areas of Belize; mycorrhizal; a symbiotic organism with trees.

Edibility: Edible; not recommended

Orange Milkycap (*Lactarius salmoneus*)

Description: A medium sized mushroom; cap is convex with a slight depression, salmon orange to whitish, with light orange tint, and exhibits concentric zones; gills are close, adnate to stalk or slightly decurrent and salmon orange in color; gills secrete salmon colored latex; stalk is short, thick and orange to pale orange; inner flesh stains reddish orange when injured.

Spore Print: Cream

Habitat & Preferred Substrate: Found in coniferous and mixed deciduous forests in areas prone to seasonal flooding; mycorrhizal; a symbiotic organism with trees.

Edibility: Edible; not recommended

Hairy Funnelcap (*Lentinus crinitus*)

Description: A small tough or rubbery fungus; cap is fuzzy, convex with a central depression and light brown to honey in color; cap fuzz fades as the mushroom is weathered, leaving only a light scalation concentrated toward the central depression; margin is rolled inwards on young fruiting bodies; gills are crowded, adnate and descending to stalk, and white, pale brown, to pale orange in color; stalk varies in thickness, is relatively short, fuzzy and white to light brown in color; annulus not present.

Spore Print: White

Habitat & Preferred Substrate: Found in open, dry habitats of Belize; commonly associated with coniferous forests though can be found in most forest types; saprophytic; grows on dead wood and roots.

Edibility: Edible; not recommended

Stinking Parasol (*Lepiota cristata*)

Description: A small to slightly medium sized mushroom with a foul odor; convex to flat with an umbonate center, white in color with reddish brown scales concentrated toward the central umbo; margin is appendiculate when young; gills are crowded, free from stalk, and white to cream in color; stalk is medium length, somewhat thick and white; annulus is usually present.
Spore Print: White; though can be green in *Lepiota. cristata* var. *viridispora*
Habitat & Preferred Substrate: Found in the soil near hardwood and conifer trees; saprotrophic; grows on organic matter.
Edibility: Inedible; possibly poisonous

Plantpot Dapperling (*Leucocoprinus birnbaumii*)

Description: A small to medium sized mushroom; Cap is egg-shaped when young, becoming bell-shaped and convex or slightly conical as it matures; cap color is bell-shaped, pale yellow with darker colored scales becoming darker toward the center; margin is closely striated in older fruiting bodies; gills are crowded, free from stalk, and yellow; stalk is medium to long with a swollen base, powdery, and yellow; annulus is sometimes present.
Spore Print: White
Habitat & Preferred Substrate: As the name suggests, this mushroom is found most commonly in plant pots; saprotrophic; grows on organic matter.
Edibility: Poisonous

White Dapperling (*Leucocoprinus cretaceus*)

Description: A relatively small mushroom; cap is egg-shaped when young, becoming convex to flat as it matures, white in color, and covered with "fluffy" to hard white scales; gills are crowded or semi-crowded, free from stalk, and white; stalk is thin, white to off-white, and features a swollen, thickly scaled base; annulus is present.
Spore Print: White to cream
Habitat & Preferred Substrate: Found mostly in plant pots; saprotrophic; grows on organic matter.
Edibility: Unknown; possibly toxic.

Fragile Dapperling (*Leucocoprinus fragilissimus*)

Description: A small mushroom; cap is bell-shaped when young, becoming convex to flat as it matures, pale yellow to white in color; distant striations run from the margin to the center of the cap; gills are close to semi-crowded, free from stalk, and white; stalk is thin, and pale yellow; annulus is present.
Spore Print: White

Habitat & Preferred Substrate: Found in moist forests and shaded areas; saprotrophic; grows on organic matter, occasionally on dung.
Edibility: Unknown

Pink Pinwheel (*Marasmius haematocephalus*)

Description: A tiny mushroom; cap is convex with a depressed center containing a small umbo; cap color is pink, fuscia to magenta with a darker center; margin is crenate with striations leading to the central umbo; gills are distant, free from stalk, and white; stalk is thin, dark reddish brown to black, with a small lighter section proximal to the cap; annulus not present.
Spore Print: White
Habitat & Preferred Substrate: Found in moist closed canopy forests of Belize; saprotrophic; found commonly on decaying leaves.
Edibility: Inedible

Pink Pinwheel (*Marasmius haematocephalus*)

Description: A small mushroom; cap is convex with a depressed center containing a small umbo; cap color is white, with an off-white to yellow center; margin is crenate with striations running from the margin to the central umbo; gills are distant, free from stalk, and white; stalk is thin, and white, annulus not present.
Spore Print: White
Habitat & Preferred Substrate: Found in moist closed canopy forests of Belize; saprotrophic; found commonly on decaying wood and moist tree stumps.
Edibility: Inedible

Orange Pinwheel (*Marasmius siccus*)

Description: A tiny mushroom; cap is convex with a depressed center containing a small umbo; cap color is pale orange to orange sometimes with a darker center; margin is crenate with striations leading to the central umbo; gills are distant, free from stalk, and white to pale yellow in color; stalk is thin, dark reddish brown to black, with a small lighter section proximal to the cap; annulus not present.
Spore Print: White
Habitat & Preferred Substrate: Found in moist closed canopy forests of Belize; saprotrophic; found commonly on decaying leaves.
Edibility: Inedible

Sullivant's Pinwheel (*Marasmius* cf. *sullivantii*)

Description: A tiny mushroom; cap is convex to flat, orange, darkening toward the center, and lacking the striations found in other species of *Marasmius*; gills are close, attached to or free from stalk, and white to pinkish white in color; stalk is thin, white proximal to the cap, darkening to orange in the center, and dark reddish brown at the base; annulus not present.

Spore Print: White

Habitat & Preferred Substrate: Found in moist closed canopy forests of Belize; saprotrophic; found commonly on decaying leaves and wood.

Edibility: Inedible

Notes: This species is phenotypically similar to M. sullivantii in its morphology; However, M. sullivantii is a species known to occur in France, therefore the 'cf.' modifier has been added.

Glowing bonnet (*Mycena margarita*)

Description: A small mushroom; cap is convex to flat with a small central depression, greyish brown in color darkening toward the center; striations run from the margin to the central depression; gills are close to semi-crowded, free from stalk, and white; stalk is thin, white and slightly translucent; annulus not present; fruiting bodies emit a faint green bioluminescent glow when dark.

Spore Print: White

Habitat & Preferred Substrate: Found in moist closed canopy forests of Belize at high elevations; saprotrophic; grows on decaying wood.

Edibility: Unknown

The Train Wrecker (*Neolentinus lepideus*)

Description: A large sized mushroom; cap is thick, convex, white to pale brown in color with flat, darker colored scales that become dense toward the center; gills are adnate, descending to stalk, and white; stalk is thick, and white to pale brown; annulus sometimes present.

Spore Print: White

Habitat & Preferred Substrate: Found mostly at the base of conifers, and less commonly at the bases of hardwoods, though can be found decaying various wooden building materials as well; saprophytic, grows on conifers and hardwoods.

Edibility: Some sources say edible, some say inedible. It's best to avoid consumption until a consensus is reached.

Common Mottlegill (*Panaeolus antillarum*)

Description: A small mushroom; cap is convex, sometimes shiny, and white, grey to golden brown; cap can be slightly wrinkled when mature; gills are close, adnate to stalk, and grey at first becoming black when mature; stalk is relatively long, thick compared to other mottlegills, and white; stalk bruises light brownish red when pinched.

Spore Print: Black

Habitat & Preferred Substrate: Found in farm areas of Belize. Coprophilous; found on animal dung

Edibility: Edible; not choice.

Banded Mottlegill (*Panaeolus cinctulus*)

Description: A small mushroom; cap is convex, becoming flat and umbonate with maturity, and light brown to tan with a dark brown band around the margin; gills are close, adnate to stalk, and cream at first becoming black when mature; stalk is relatively long, thin, and white to light brown; stalk bruises brownish red when pinched.

Spore Print: Black

Habitat & Preferred Substrate: Found in farm areas of Belize. Coprophilous; found on animal dung

Edibility: Psychoactive

Blue Meany (*Panaeolus cyanescens*)

Description: A small mushroom; cap is convex, and white to light brown; gills are close, adnate to stalk, and grey at first becoming black when mature; stalk is relatively long, and white; stalk bruises blue when pinched; cap stains bluish black.

Spore Print: Black

Habitat & Preferred Substrate: Found in farm areas of Belize. Coprophilous; found on animal dung

Edibility: Psychoactive

Petticoat Mottlegill (*Panaeolus papilionaceus*)

Description: A small mushroom; cap is convex to conical, and light brown; margin has a white or light colored rim; gills are close, adnate to stalk, and grey at first becoming black when mature; stalk is relatively long, and brown to reddish brown.

Spore Print: Black

Habitat & Preferred Substrate: Found in farm areas of Belize. Coprophilous; found on animal dung

Edibility: Edible; not choice

Shaggy Scalycap (*Pholiota squarrosa*)

Description: A medium sized mushroom; cap is bell-shaped, convex, to flat as it matures and is covered in thick spiky scales; cap is brown to light brown in color, often getting lighter toward the margin; gills are semi-crowded, adnate to stalk, off-white, turning olive green to brownish in color when mature, and covered by a veil when young; stalk is thick, covered in spiky scales, and light brown in color, becoming lighter proximal to the cap; annulus is present after veil breaks.

Spore Print: Cinnamon to rusty brown

Habitat & Preferred Substrate: Found in hardwood forests of Belize; parasitic; causes white rot pathogen in living trees.

Edibility: Inedible. Poisonous when consumed with alcohol.

Common Brittlestem (*Psathyrella candolleana*)

Description: A small mushroom; cap is conical, convex, to flat with maturity; cap is honey colored, becoming lighter proximal to the margin with specks of white veil remnants when young; when mature and dry cap can appear whitish; gills are crowded, notched or free from stalk, white at first and dark brown when mature; stalk is short, hollow, and white; annulus rarely present.

Spore Print: Dark purplish brown

Habitat & Preferred Substrate: Found in tall grassy areas of Belize; saprotrophic; grows on organic matter

Edibility: Edible, though not recommended for consumption as there are poisonous and inedible lookalikes.

Hairy Brittlestem (*Psathyrella hirta*)

Description: A tiny mushroom; cap is convex to semi-conical, brown in color, covered with fuzzy white tufts of hair; gills are semi-close, adnate to stalk, and dark brown or black; stalk is short, hollow, and white with short white hairs; annulus not present.

Spore Print: Black

Habitat & Preferred Substrate: Found in moist, dark locations; saprotrophic; grows on organic matter, usually soil and rotting wood.
Edibility: Unknown

Magic Mushroom (*Psilocybe cubensis*)
Description: A small to medium sized mushroom; cap is convex to umbonate, cinnamon brown to golden brown, becoming lighter proximal to the margin; white, speckled veil remnants can be seen on most young fruiting bodies and some mature fruiting bodies; gills are close to semi-crowded; adnate or adnexed to stalk, and purplish to black in color; stalk is thick, short to medium length, white to yellowish; annulus is present; the flesh of this mushroom stains blue when injured or bruised.
Spore Print: Dark purplish brown
Habitat & Preferred Substrate: Found in farm areas of Belize. Coprophilous; found on animal dung
Edibility: Psychoactive

Golden Brittlegill (*Russula flavida*)
Description: A medium sized mushroom; cap is convex to flat, and yellow; gills are close to crowded, and white to pale yellow; stalk is smooth, medium length, thick and yellow
Spore Print: Pale yellow
Habitat & Preferred Substrate: Found in moist forests of Belize; mycorrhizal; a symbiotic organism with hardwoods
Edibility: Unknown

Boletes

Butterfoot Bolete (*Aureboletus/Boletus auriporus*)

Description: A medium to large sized mushroom with a spongy, thick, convex to flat, light reddish-brown cap; pore surface features spongy, yellow tubes, turning reddish brown when mature; stalk is thick, and reddish brown, becoming a cream-yellow color proximal to the pore surface, and featuring faint yellow fibrous striations; annulus not present; cap flesh is off-white; flesh does not stain blue when injured or bruised.

Spore Print: Brown

Habitat & Preferred Substrate: Found in open and partly shaded areas of Belize in soil near trees; mycorrhizal; a symbiotic organism with hardwoods.

Edibility: Edible

Pineapple Bolete (*Boletellus ananas*)

Description: Medium sized fruiting body; cap round to convex, and covered in large pink to grey wooly scales; pore surface features spongy, yellow tubes which bruise dark blue to reddish brown when pressed; stipe is a medium length and relatively thick, and smooth; stipe pink to pale yellow in color; ascending annulus not present in mature fruiting bodies; flesh is initially white, staining blue when injured.

Spore print: Brown

Habitat & Preferred Substrate: found in broadleaf and coniferous forests of Belize. Ecology of this mushroom is unknown; associated with pine trees.

Edibility: Not recommended

Similar Species: Old Man of the Woods (*Strobilomyces* sp.)

Dupain's Bolete *(Boletus/Rubroboletus dupainii)*

Description: A medium sized mushroom with a sticky, convex, red to burgundy colored cap; pore surface features spongy, reddish orange tubes, turning dark orange to brown when mature; stalk is thick, yellow to orange, and becomes a reddish toward the base; annulus not present; cap, pore surface, and inner flesh stain dark blue when bruised or injured.

Spore Print: Dark brown

Habitat & Preferred Substrate: Found in open and partly shaded areas of Belize in soil near trees; mycorrhizal; a symbiotic organism with hardwoods.

Edibility: Edible. However, this mushroom can be confused with poisonous boletes. For this reason, beginning foragers should avoid consuming this mushroom.

Cracked Bolete (*Boletus variipes* var. *fagicola*)

Description: A medium to large sized mushroom with a spongy, thick, convex, dark brown cap; cap surface appears dry and cracked when mature; pore surface features spongy, white to pale yellow tubes; stalk is very thick, reticulated, and white to light brown; annulus not present; cap, inner flesh, and pore surface do not stain when injured or bruised.
Spore Print: Brown
Habitat & Preferred Substrate: Found in open and partly shaded areas of Belize in soil near trees; mycorrhizal; a symbiotic organism with hardwoods.
Edibility: Edible

Wrinkled Bolete (*Leccinum viscosum*)

Description: A medium to large sized mushroom with a sticky, thick, convex, yellow, tan to brown cap; cap surface appears wrinkled when mature; pore surface features spongy, yellow to brownish tubes; stalk is thick, and white to yellowish, featuring reddish brown edged scabers; annulus not present; inner flesh of cap and stalk stain pink to reddish brown when injured.
Spore Print: Brown
Habitat & Preferred Substrate: Found in Mountain Pine Ridge, Belize in soil near pine trees; namely *Pinus caribeae* and *Quercus peducularis* (Ortiz-Santana & Halling 2009); Likely mycorrhizal and symbiotic with pine trees.
Edibility: Unknown
Note: The photos provided are of *Leccinum rugiosiceps*, which is visually identical, due to lack of photos of the Belizean variety *B. viscosum*. This mushroom can still be identified using the photos; however, accurate identification must be made using the description provided.

Powdery Sulfer Bolete (*Pulveroboletus ravenelii*)

Description: A medium to relatively large sized mushroom; cap is dry or sticky, thick, convex, and reddish orange; cap is encompassed by a yellow, powdery partial veil when young; pore surface features spongy white, yellow to yellowish brown tubes; stalk is a medium length and thickness, yellow to white in color, and sometimes features grooved striations; annulus not present; cap and pore surface stain blue when bruised or injured; stalk stains pink when cut.

Spore print: Greenish brown

Habitat & Preferred Substrate: Found in open and partly shaded areas of Belize in soil near trees; mycorrhizal; a symbiotic organism with conifers.

Edibility: Edible

Stinkhorns

Latticed Stinkhorn (*Clathrus crispus*)

Description: A small to medium sized stinkhorn; fruiting bodies emerge from white egg-like volva casing; fruiting body color is pinkish orange, turning lighter toward the base, and shape is oblong or spherical with a distinct, lattice-shaped structure. *C. crispus* can be told apart from *C. ruber* by the lattice structure, *C. crispus* has circular latticing, whereas *C. ruber* latticing is irregular.
Spore Print: Brown
Habitat & Preferred Substrate: Found in moist, closed canopy forests, and sometimes tall grassy areas; saprotrophic; grows on organic matter, usually rotting chunks of wood.
Edibility: Inedible and potentially poisonous

Lantern Stinkhorn (*Laternea triscapa*)

Description: A relatively small, oddly shaped stinkhorn; fruiting bodies emerge from white egg-like volva casing; two or more orange to red column extensions project upward and fused together at the top; the inner surface of the top is usually coated with a brownish slime which emits a putrid smell;
Spore Print: Dark brown slime
Habitat & Preferred Substrate: Found in moist, closed canopy forests, and sometimes tall grassy areas; saprotrophic; grows on organic matter, usually rotting chunks of wood.
Edibility: Poisonous
Similar Species: Laternea pusilla

Veiled Stinkhorn (*Phallus indusiatus*)

Description: A medium to large sized stinkhorn mushroom; fruiting body is phallic shaped, with a dark brown or greyish cap, and a thick, white stalk; cap is smooth to ridged with an opening at the top, and has a net-shaped "skirt" extending from the margin and draping to the ground around the stalk.
Spore Print: Yellowish brown
Habitat & Preferred Substrate: Found on the ground in moist forest areas; saprotrophic; grows on organic matter.
Edibility: Edible

Cup, Jelly, & Bird's Nest Fungi

Jelly Ear Fungus (*Auricularia auricula-judae*)

Description: A relatively small, but conglomerate fungus; as the common name suggests, fruiting bodies are soft or tough (depending on maturity) and gelatinous, resembling a cup-shaped, wrinkled ear; fruiting body is attached to substrate and stalk typically isn't present; color is tan brown, purple or reddish; specimens can be smooth or have a short, velvety fuzz; the inner portion is usually a lighter color and has a vein-like texture.

Spore Print: White to cream-yellow

Habitat & Preferred Substrate: Found in clusters around dense, moist, closed canopy forest areas of Belize often near water; saprophytic/parasitic; grows on dead and living trees.

Edibility: Edible

Tripe Fungus (*Auricularia mesenterica*)

Description: A small to medium sized bracket-like, gelatinous fungus; fruiting bodies are wavy and lobed with cortical surface zonations; fruiting bodies are attached to substrate; upper surface may have a short, velvety fuzz and color is grey, tan or purplish in appearance; pore surface is smooth and whitish but becomes reddish brown when mature; inner flesh is tough and rubbery.

Spore Print: White

Habitat & Preferred Substrate: Found in moist forest areas of Belize; saprophytic; found on dead and rotting wood

Edibility: Inedible

Fan-shaped Jelly Fungus (*Dacryopinax spathularia*)

Description: A tiny fungus; as the common name suggests, fruiting bodies are fan-shaped and gelatinous in texture; appear dark yellow to orange in color; surface is smooth; stalk is short and darker than upper portion.

Spore Print: White

Habitat & Preferred Substrate: Widely distributed around Belize; saprotrophic; can grow on a vast variety of organic matter.

Edibility: Edible

Snow Fungus (*Tremella fuciformis*)
Description: a gelatinous fungus; fruiting body is lobed, viscid and translucent white, occasionally with off-white patches when weathered.
Spore Print: White
Habitat & Preferred Substrate: Found on branches in moist forests of Belize; suspected to be parasitic or saprophytic; grows on tree live and dead branches.
Edibility: Edible

Cloud Ear Fungus (*Auricularia nigricans*)
Description: A relatively small, thin cup/ear shaped fungus; upper surface features a velvety fuzz which **is** brown, to white, to light brown in color as it matures; inner-surface is smooth and rubbery and dark brown to magenta in color, turning dark brown or black as it ages; fruiting bodies are attached to substrate, no stalk present.
Spore Print: White
Habitat & Preferred Substrate: Found on hardwood trees in moist forest areas of Belize; saprophytic; found on dead and rotting wood
Edibility: Edible

Pink Goblet (*Cookeina speciosa*)
Description: A small, thin and rubbery mushroom; cap is cup shaped, and light orange to pink in color; outer cup surface tends to be a lighter color than the darker, smooth inner-surface when mature, and has tiny hairs concentrated around the upper margin; stalk is thin and white to pale yellow in color.
Spore Print: White
Habitat & Preferred Substrate: Found in moist, closed canopy forest areas of Belize; usually on the ground, clustered on damp, broken pieces of hardwoods; saprophytic; found on dead and rotting wood.
Edibility: Edible; not choice. Best to avoid consumption until further research has been carried out.

White Eyelash Cup (*Cookeina tricholoma*)
Description: A small, thin and rubbery mushroom; cap is cup shaped, and light orange to pink in color; outer cup surface tends to be a lighter color than the darker, smooth inner-surface when mature, and has elongated hairs

concentrated around the upper margin; stalk is thin, short and white to pale yellow in color.
Spore Print: Pink ascospores
Habitat & Preferred Substrate: Found in moist forests with heavy canopy; saprotrophic; grows on organic matter, usually chunk of decaying wood.
Edibility: Edible; not choice. Best to avoid consumption until further research has been carried out.
Similar Species: *Scutellinia scutellata* appears similar, though has dark colored hairs instead of white.

Barnyard Cup (*Peziza vesiculosa*)

Description: A relatively small, cup-shaped fungus; fruiting bodies are usually clustered; outer surface is pale brown to pale yellow with a darker inner surface; attached to substrate, stalk not present.
Spore Print: White
Habitat & Preferred Substrate: Found in farm areas of Belize. Coprophilous; found on animal dung
Edibility: Poisonous

Tropical Bird's Nest Fungi (*Cyathus berkeleyanus*)

Description: A tiny, clustering fungus; fruiting bodies are cup-shaped and resemble a bird's nest; young fruiting bodies appear fuzzy; outer surface is brown; inner surface is grey, smooth, and somewhat shiny with distant grooves occasionally present; spores are located at the bottom of the nest, grey to black in color, and egg shaped.
Spore Print: No spore print; spores are egg shaped portion of the nest.
Habitat & Preferred Substrate: Found in moist areas of Belize; saprotrophic; grows on organic matter, mostly on chunks of rotting wood.
Edibility: Inedible

Coral & Club Fungi

Cordycepioid Fungi (*Akanthomyces/Ophiocoordyceps* sp.)

Description: Cordycepioid fungi are parasitic fungi that affect arthropods, such as moths, ants, spiders and grasshoppers; fruiting bodies protruding from the arthropod are thin, sometimes in spiky clusters, occasionally with a single protrusion featuring a small bulbous tip, and are white to yellowish in color.

Spore Print: White

Habitat & Preferred Substrate: Parasitic; found protruding from arthropod hosts, typically in dense, closed canopy forests.

Edibility: Inedible/unknown

Note: The fungi species on the left, *Akanthomyces tuberculata* typically affects moths; the photo on the right is of the *Ophiocordyceps unilateralis* group, which affects ants.

Blue-tipped Coral (*Phaeoclavulina cyanocephala*)

Description: A relatively large coral fungus; as the name suggests, fruiting bodies are coral-shaped, and have blunt, greenish blue tips; branching projections are tan to light brown; blue tips are sometimes not present in mature or weathered specimens; branches bruises brown when injured or bruised.

Spore Print: Brownish Yellow

Habitat & Preferred Substrate: Found in closed canopy forest areas; saprophytic; grows at the base of trees.

Edibility: Unknown

Thin-branch Coral (*Lentaria surculus*)

Description: A medium to large sized coral fungus; fruiting body has at a thick off-white to brownish stalk; branches extend upward and are off-white, yellowish, to light brown in color; branches end in highlighted whitish tips, with ~1-3 small, projected points.

Spore Print: White

Habitat & Preferred Substrate: Found moist semi-closed, to closed canopy forests, in close proximity to trees; mycorrhizal; a symbiotic organism with vascular plants.

Edibility: Inedible

King Alfred's Cake (*Daldinia concentrica*)
Description: A bulbous or irregular lump of a tree bracket with multiple inner layers; fruiting body is black to dark purplish brown in color;
Spore Print: Black
Habitat & Preferred Substrate: Found on the sides of some trees in dryer areas of Belize; saprophytic; mostly grows on ash trees (*Fraxinus* sp.).
Edibility: Inedible

Candlesnuff Fungus (*Xylaria hypoxylon*)
Description: A small fungus, typically found in groups; fruiting bodies are thin, finger-like projections with antler shaped extensions protruding from the top of some specimens; color of fruiting bodies ranges from white with a powdery texture, to black with a solid texture depending on maturity; white fruiting bodies usually have a black base.
Spore Print: Black
Habitat & Preferred Substrate: Found commonly in moist forest of Belize; saprotrophic; grows on organic matter, usually rotting wood
Edibility: Inedible

Dead Man's Finger (*Xylaria polymorpha*)
Description: A medium sized fungus, typically found in groups or clusters; fruiting bodies are thick, finger-like, have a tough texture, and are often fused together near the base; fruiting bodies appear black with white tips when young, and greyish bluish grey and black when mature.
Spore Print: Black
Habitat & Preferred Substrate: Found commonly in moist forest of Belize; saprotrophic; grows on organic matter, usually rotting wood
Edibility: Inedible

Puffballs & Earthstars

Spiny Puffball (*Lycoperdon echinatum*)

Description: A small puffball mushroom; fruiting body is spherical and covered in tough spiky projections; spikes vary in size depending on maturity; color is white to off-white when young, turning brown when mature; inner flesh is white and firm to spongy when young, and dark brown and moist or powdery as it ages.
Spore Print: Brown
Habitat & Preferred Substrate: Found in pastures, forests and gardens; saprotrophic; found growing in soil or woody debris.
Edibility: Edible when young and inner flesh is still solid and white; not recommended.

Common Puffball (*Lycoperdon perlatum*)

Description: A small puffball mushroom; fruiting body is spherical or bulbous with a thinner base and is covered fine dotted bumps that fade when mature; color is white, grey or brown depending on maturity; fruiting body has a hole in the center for spore release that may not be present in younger specimens; inner flesh is white and firm to spongy when young, and dark brown and moist or powdery as it ages.
Spore Print: Brown
Habitat & Preferred Substrate: Found in pastures, forests and gardens; saprotrophic; found in organic matter on the ground.
Edibility: Edible when young and inner flesh is still solid and white.

Stump Puffball (*Lycoperdon* cf. *pyriform*)

Description: A small puffball mushroom; fruiting body is spherical or bulbous with a thinner base and is covered fine dotted bumps that fade when mature; color is white, pale brown, reddish brown or blackish brown depending on maturity; fruiting body has a hole in the center for spore release that may not be present in younger specimens; inner flesh is white and firm to spongy when young, and dark brown and moist or powdery as it ages.
Spore Print: Brown
Habitat & Preferred Substrate: Found in pastures, forests and gardens; saprophytic; found growing on dead and rotting wood
Edibility: Edible when young and inner flesh is still solid and white.

Similar Species: This species may look identical to *Lycoperdon perlatum*. The primary difference between the two is that *L. pyriforme* is more often found growing on wood, whereas *L. perlatum* is found growing on the ground.

Dead Man's Foot (*Pisolithus arhizus*)

Description: A medium to large sized fungus; texture is tough and crusty, spherical when young, and irregularly shaped with a swollen top when mature; color is light yellowish brown, to dark brown with black spots when mature.
Spore Print: Reddish brown
Habitat & Preferred Substrate: Found commonly in human disturbed areas; mycorrhizal; a symbiotic organism with vascular plants.
Edibility: Inedible

Rounded Earthstar (*Gaestrum saccatum*)

Description: A small to medium sized earthstar fungus; young fruiting bodies resemble a puffball with a pointy center; mature fruiting body is grey to brownish with a spherical center, and an outer layer resembling triangular flower pedals; at the center of the sphere is a discolored, circular area with a hole for spore release; pedal-like extensions are usually a different color than the spherical center, ranging from grey, brown to reddish brown, and furl downwards when mature.
Spore Print: Brown
Habitat & Preferred Substrate: Found in the soil underneath trees, usually in close proximity to water; saprobic; found growing on organic matter.
Edibility: Inedible

Berkley's Earthstar (*Geastrum berkeleyi*)

Description: A small to medium size, erect, earthstar fungus; fruiting body is comprised of a spherical to oblate head, elevated on a short stalk over a secondary layer that resembles flower pedals; the spherical head portion is dark greyish green to light brown when dry with a pointed central hole, surrounded by small notches; the pedal-looking layer curves downward, and is off-white, turning reddish brown when mature.
Spore Print: Brown
Habitat & Preferred Substrate: Uncommon; saprotrophic; found growing on organic matter near *Fraxinus* sp.
Edibility: Inedible

Slime Molds

Note: Slime molds aren't actually molds, nor are they fungi; they are an aggregate of single celled organisms, similar to the Siphonophores of the ocean. However, since they are commonly mistaken for fungi, I have added a few common species you may see around. Slime mods are not edible. Don't try to eat them.

Carnival Candy Slime Mold (*Arcyria denudate*)

Description: A conglomerate mass of small reddish to pink fruiting bodies; tubular to oblong in shape, sometimes with short, visible stalks; immature fruiting bodies are white and round.
Preferred Substrate: Decaying wood

Bell-shaped Slime Mold (*Physarella oblonga*)

Description: a mass of tiny fruiting bodies; upper spore-bearing portion (sporangium) is grey to light yellowish brown, bell shaped when young, and flower shaped when mature; stalk is long compared to sporangium and red to dark brown in color.
Preferred Substrate: Live or dead plants; fruiting bodies in photos found on live orchid leaves and mulch.

Chocolate Tube Slime (*Stemonitis splendens*)

Description: small tuft-like fruiting body conglomerates; individual fruiting bodies are tubular in shape, and chocolate brown in color; clusters usually have a powder of brown spores beneath them.
Preferred Substrate: Decaying wood and lumber

5 COMMON RULES WHEN FORAGING FOR WILD MUSHROOMS

Rule # 1.

Never eat a mushroom that has not been positively identified. Use all ID factors (shapes, texture, gill/tube/pore/spine/ridge type, spore print color, spore shape, etc.) for identification. Ask an expert for a second opinion.

Rule # 2.

When foraging for edible mushrooms as a beginner, it is best to avoid mushrooms with gills. The most easily identifiable edibles in Belize are boletes (avoid blue staining and red colored boletes), oyster mushrooms, and puffballs that are still young and white on the inside. Avoid oyster mushrooms on coniferous trees. They may be angel wings

Rule # 3.

Never eat wild mushrooms raw!

Rule # 4.

When a mushroom looks rotten, old or smells putrid it is no longer edible. If you eat it, you will get sick. Simple as that!

Rule # 5.

When transporting or keeping mushrooms its best to use a basket instead of a plastic bag or closed container, mushrooms need to breath and when you keep them in an airtight environment, they will spoil quickly.

Section VII

CITIZEN SCIENCE

A Beginner's Guide to Field Mycology

21ST CENTURY EXPLORER

Belize is a country with enormous biodiversity; this holds true with plants, animals, and fungi alike. Due to the myriad of species that can be found within the nation, there are still many out there which are undiscovered and undescribed. That's where you come in!

Citizen Science is the name given to the non-scientist contributors of scientific data records. In many cases, especially when it comes to biology, citizen scientists outnumber the actual scientists currently in a particular field. As a result of this, these contributors are an extremely valuable asset to each field. Mycology is one of the fields of study that simply doesn't get enough attention when it comes to species diversity. This is a shame because fungi are fascinating; and I truly believe that we just need to pause and observe our surroundings while we're outdoors to find that there is this spectacular hidden world all around us, and beneath our feet.

The next few pages will be dedicated to informing you, the up-and-coming mycologist, about the different tools of the trade. The common techniques I will be reviewing include photo documentation, chemical testing, and proper data collection.

Please note that protected areas such as National Parks and Wildlife Sanctuaries of Belize may require proper permissions to study, collect or test mushrooms. Please make sure you are aware of the restrictions which apply to any non-public locations. Contact the Belize Forest Department if you're unsure.

USEFUL TOOLS FOR MYCOLOGICAL FIELD RESEARCH

Weave hats are not only useful for sun protection; they also double as a basket for transporting mushrooms

Chemical tests are used to identify species and to elucidate the presence of some toxins

Always keep a field notebook on you to record your finds and interesting information

Smart phones are useful for their cameras and array of helpful apps in the field. Make sure to download offline maps for your area so you don't get lost

iNaturalist

A useful app that assists with identifications of plants, animals, and fungi

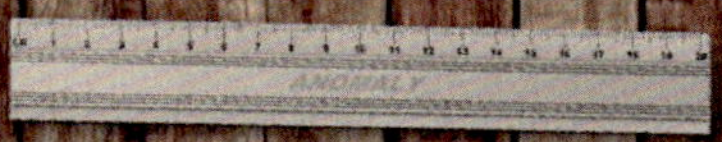

Rulers are a helpful tool to add into identification photos

PROPER PHOTO DOCUMENTATION

Photo documentation is important when trying to identify fungi, especially for beginners. Whether you're comparing photos to those found in this guide, sharing with community based identification groups, or sending to an expert, there are crucial details that should be exhibited in the images you collect.

In-situ and habitat shots

Mushrooms growing from a dead tree stump in a deciduous forest.

When taking photos of fungi for identification, it's always good to first take a photo of the fruiting body in-situ (in its natural setting). Try to capture the tree or substrate where it's growing in the photo. If the fruiting body is emerging from the ground, simply take a photo of the habitat surrounding the fungi as an addition to your documentation. It is good practice to carry a small ruler or measuring stick with you when mushroom hunting. Set the measuring tool in front of the mushroom for in-situ shots to provide others with a general idea of their size.

Standard Specimen Photos

Jackson's Slender Caesar Amanita jacksonii
Photo by Alan Rockefeller

Please note that the mushroom is simply the fruiting body of a much larger organism – the mycelium. Plucking a mushroom from its substrate is the equivalent of pulling a fruit from a tree. You will not being harming the organism, and it will be able to reproduce just fine. If anything, you will be acting as a mechanism of spore dispersal.

Now that you have the in situ and habitat shots, you can remove the mushroom from its place, either with a knife or by hand. Once the mushroom is removed, take close-up photos of the full fruiting body, as well as the gills or pore surface on the underside. If the fungus you're photographic is something like a cup fungus, which has neither gills nor a pore surface, the fruiting body close-up alone is sufficient. If there are multiple stages of growth (young and old mushrooms), present them in the photo.

Staining & Bruising

Blue staining on a cross sectioned bolete
Photo by Alan Rockefeller

Once the fruiting body shots are complete, use your hands, a knife, or something with an edge to cut the mushroom in a cross-section, wait to see if there is any staining or bruising and take a photo. If the mushroom is too small or brittle for a cross section, pinch the stem and scar or lightly bruise the cap or pore surface to check for staining. For example, *Agaricus semotus* will stain yellow when cut or bruised, *Psilocybe cubensis* will stain blue when the stem is pinched, and the pore surface of some boletes will bruise blue or brown when pressed. Another valuable tool for fungi identification is chemical testing. Chemical reagents can be used on the cap, stalk, gills, pores, spores and flesh of a mushroom to differentiate species that may otherwise appear identical. Take a photo of any reactions that occur!

Spore Printing

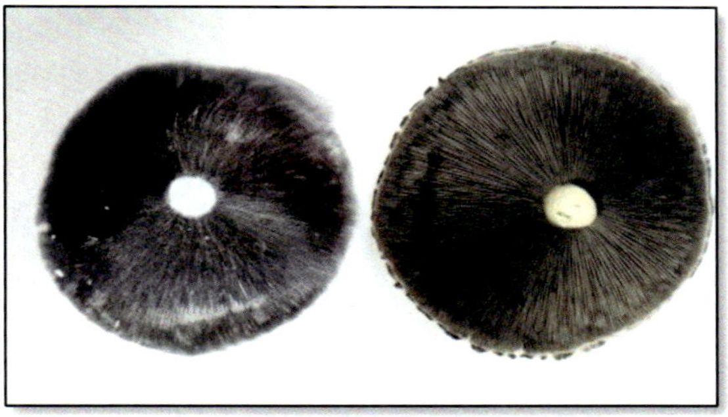

Spore print on white paper
Photo by Suhaib Firdous Yatoo

Some species can be easily determined by spore prints. For example, the edible Parasol Mushroom (*Macrolepiota procera*) looks nearly identical to the poisonous false parasol A.K.A. "The Vomiter" (*Chlorophyllum molybdites*). However, through spore printing, one can easily determine which is which. The spore color of *C. molybdites* is green; the true parasol mushroom has white colored spores.

In order to obtain a spore print, you must remove the cap of a mushroom, and place it on a sheet of paper gills down. I usually prefer using tinfoil in place of paper as the spores of some mushrooms can be light colored or white and may not be visible. Once you've placed it on your preferred surface, cover it with a bowl overnight. In the morning, remove the bowl and lift the mushroom to see the print. If the mushroom doesn't release spores after the first night, give it more time. Once you've obtained the spore print, take a clear photo of the spores next to the cap of the mushroom.

CHEMICAL TESTING

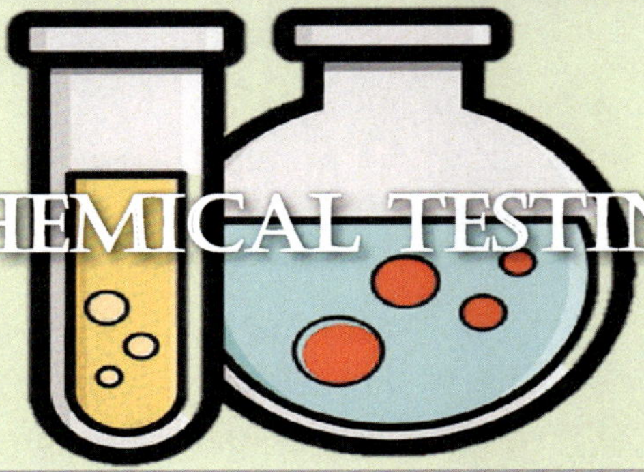

Various chemical tests can be used in the identification of mushrooms. Chemicals often used in the field by mycologists include: ammonia, ferrous sulfate, potassium hydroxide, vanillin (sulfo-vanillin), sulfuric acid, hydrochloric acid and Iodine. Different chemicals are used for different things, though the expected reaction will always be a color change. Some color change reactions can persist, though some can occur and disappear instantaneously, some exhibit multiple colors. It is important to note the reactions that took place when using the chemical testing method, and take photos if possible. Many fungi in biodiverse locations like Belize have not had extensive chemical testing done, so you may find something new and interesting! If you are planning on trying this method, please note that you will also be doing it to collect new and useful data for mycological science, and not exclusively for identification purposes. Good luck out there!

General information on mushrooms targeted with each chemical compound is based on the article “Testing Chemical Reactions” found on the Mushroom Expert website (Kuo, 2006).

Warning: Do not consume any mushrooms on which you have used chemical reagents.

Potassium hydroxide (KOH) 5%

Potassium hydroxide is probably the most commonly used chemical reagent of mycologists. This chemical compound is typically used in the identification of a wide range of fungi including polypore, gilled mushrooms and boletes. With polypores, add a drop to the cap or inner flesh; for gilled mushrooms add a drop to the cap; and for boletes, just as with the other two reagents, add a drop to the cap, stalk, pore surface or inner flesh. Record any reactions that take place

Green satins from KOH on Gymnopus subsect. Alkalivirentes
Photo by Alan Rockefeller

Ammonia (NH_3)

Ammonia is commonly used on species of bolete mushrooms. Mushrooms in the Boletaceae family can be identified by their typical mushroom shape, and a spongy pore surface comprised of tubes instead of gills. There are at least 52 species of bolete in Belize (Ortiz-Santana et al., 2007). When using this chemical reagent, place a drop on different parts of the mushroom including the cap, stalk, pore surface and inner flesh and record any reactions.

Orange stains from NH3 on cap of Boletus subvelutipes by ammonia
Photo by Damon Brunette

Ferrous sulfate ($FeSO_4$) 10%

Ferrous sulfate is an iron salt commonly used to identify species of bolete and brittlegills (*Russula* sp.). As with ammonia, a drop can be added to the cap, stalk, pore surface or inner flesh of boletes to trigger a color change reaction. For brittlegills, simply add a drop to the stalk. Record any reactions that take place.

Dark stains from FeSO4 on Leccinum subleucophaeum
Photo by Mike Hopping

Chemicals reagents bottles can be easily transported in mint cans
Photo by Alan Rockefeller

EFFICIENT DATA COLLECTION

Data collection is very important to accumulate a better understanding of species and how they function. You can do this in just about any outdoor situation. Community-based apps/websites like iNaturalist and Mushroom Observer make data collection a convenient, fun and educational experience. However, I understand not everyone has a smartphone or laptop, and in my own personal opinion it's advantageous to keep paper records.

Alongside everything else that has been covered in this book, it's important to realize when taking data that there are factors in determining the identity of fungi other than texture and appearance. Scent and even taste of mushrooms can be useful as well. In this section I will explain.

Taste and Spit

While exploring the possibilities of mycological data collection, you may come across some who use the term "taste and spit test". This test requires the observer to take a small piece of the mushroom into their mouths and quickly spit it out to determine the taste. Under no circumstances should the observer swallow the piece of mushroom when doing this test, and for this risk alone, beginners should not use this method for identification. Both taste and smell can sometimes determine the species being observed. For example, a brittle gill (*Russula* sp.) with a red cap could be edible, or could be the poisonous *Russula emetica.* Using the taste and spit test, *R. emetica* will typically have a spicy taste to it; the edible *Russula* will have a mild taste. Again, this test is not recommended for beginners.

Sense of Smell

Fungi identification can really be an immersive experience of the senses. The color, texture, taste and smell of different species can be distinctive. Sometimes only one or two senses are required to determine an ID; sometimes all of them are. However, each sense has an importance in the process. A textbook example of narrowing down identification via smell would be the distinct odor of the stinkhorn family, Phallaceae. Stinkhorns smell like rotting meat, and because of this, they are some of the easiest to fungi to identify. The reason for their putrid smell is spore dispersal. The stench from stinkhorn fungi attracts flies, which think there is something dead for them to feed on. When a fly lands on the fungus, sticky spores attach to its legs and are carried with it to other locations. Other mushroom smells include: mealy, fishy, anise-like, bleach-like, garlicy, and most commonly, n discernable odor at all (Kuo, 2006). Try it sometime; maybe you'll enjoy the smells of the fungi kingdom!

Put your knowledge to the test!

On the following pages, you will find a few pre-made datasheets. I challenge all who read this book to explore Belize; go hiking, camping, or adventuring. When you find a mushroom, try using what you've learned and put it into practice. Take photos, check for reactions, or just try your best to guess the identity based on what you know. Have fun, be safe, and stay curious!

Thank you all for reading!

Date:______________

Habitat: (Deciduous Forest/Coniferous Forest/Human Disturbed)

Habitat Notes (explain your surroundings, substrate type, etc.)

Fungi Type: (Mushroom/Bracket/Puffball/Other)

Underside: (Gills/Ridges/Tubes/Pores/Toothed/Other)

Cap Type: (Flat/Umbonate/Convex/Funnel Shaped/Depressed/Umbilicate/Uplifted)

Gill Distance: (Distant/Close/Crowded)

Gill Type: (Free/Adnexed/Notched/Adnate/Decurrent)

Annulus Present? (Yes/No)

Does Stain When Bruised or Injured? (Yes/No)

Stain Color?

____

Chemical Reagent used? (Ammonia/$FeSO_4$/KOH)

Describe Reaction:____________________________

____

Photos Taken? (Yes/No)

Spore Color?

____

Suspected ID:______________________________

Notes

____

____

____

Date:______________

Habitat: (Deciduous Forest/Coniferous Forest/Human Disturbed)

Habitat Notes (explain your surroundings, substrate type, etc.)

Fungi Type: (Mushroom/Bracket/Puffball/Other)

Underside: (Gills/Ridges/Tubes/Pores/Toothed/Other)

Cap Type: (Flat/Umbonate/Convex/Funnel Shaped/Depressed/Umbilicate/Uplifted)

Gill Distance: (Distant/Close/Crowded)

Gill Type: (Free/Adnexed/Notched/Adnate/Decurrent)

Annulus Present? (Yes/No)

Does Stain When Bruised or Injured? (Yes/No)

Stain Color?

____

Chemical Reagent used? (Ammonia/$FeSO_4$/KOH)

Describe Reaction:____________________________

____

Photos Taken? (Yes/No)

Spore Color?

____

Suspected ID:______________________________

Notes

____

____

____

Date:______________

Habitat: (Deciduous Forest/Coniferous Forest/Human Disturbed)

Habitat Notes (explain your surroundings, substrate type, etc.)

Fungi Type: (Mushroom/Bracket/Puffball/Other)

Underside: (Gills/Ridges/Tubes/Pores/Toothed/Other)

Cap Type: (Flat/Umbonate/Convex/Funnel Shaped/Depressed/Umbilicate/Uplifted)

Gill Distance: (Distant/Close/Crowded)

Gill Type: (Free/Adnexed/Notched/Adnate/Decurrent)

Annulus Present? (Yes/No)

Does Stain When Bruised or Injured? (Yes/No)

Stain Color?

____

Chemical Reagent used? (Ammonia/$FeSO_4$/KOH)

Describe Reaction:__________________________

____

Photos Taken? (Yes/No)

Spore Color?

____

Suspected ID:____________________________

Notes

____

____

____

Date:______________

Habitat: (Deciduous Forest/Coniferous Forest/Human Disturbed)

Habitat Notes (explain your surroundings, substrate type, etc.)

Fungi Type: (Mushroom/Bracket/Puffball/Other)

Underside: (Gills/Ridges/Tubes/Pores/Toothed/Other)

Cap Type: (Flat/Umbonate/Convex/Funnel Shaped/Depressed/Umbilicate/Uplifted)

Gill Distance: (Distant/Close/Crowded)

Gill Type: (Free/Adnexed/Notched/Adnate/Decurrent)

Annulus Present? (Yes/No)

Does Stain When Bruised or Injured? (Yes/No)

Stain Color?

____

Chemical Reagent used? (Ammonia/$FeSO_4$/KOH)

Describe Reaction:__________________________

____

Photos Taken? (Yes/No)

Spore Color?

____

Suspected ID:____________________________

Notes

____

____

____

Date:______________

Habitat: (Deciduous Forest/Coniferous Forest/Human Disturbed)

Habitat Notes (explain your surroundings, substrate type, etc.)

Fungi Type:
(Mushroom/Bracket/Puffball/Other)

Underside:
(Gills/Ridges/Tubes/Pores/Toothed/Other)

Cap Type: (Flat/Umbonate/Convex/Funnel Shaped/Depressed/Umbilicate/Uplifted)

Gill Distance: (Distant/Close/Crowded)

Gill Type:
(Free/Adnexed/Notched/Adnate/Decurrent)

Annulus Present? (Yes/No)

Does Stain When Bruised or Injured? (Yes/No)

Stain Color?

____

Chemical Reagent used?
(Ammonia/$FeSO_4$/KOH)

Describe Reaction:________________________

____

Photos Taken? (Yes/No)

Spore Color?

____

Suspected ID:__________________________

Notes

____

____

____

Date:______________

Habitat: (Deciduous Forest/Coniferous Forest/Human Disturbed)

Habitat Notes (explain your surroundings, substrate type, etc.)

Fungi Type:
(Mushroom/Bracket/Puffball/Other)

Underside:
(Gills/Ridges/Tubes/Pores/Toothed/Other)

Cap Type: (Flat/Umbonate/Convex/Funnel Shaped/Depressed/Umbilicate/Uplifted)

Gill Distance: (Distant/Close/Crowded)

Gill Type:
(Free/Adnexed/Notched/Adnate/Decurrent)

Annulus Present? (Yes/No)

Does Stain When Bruised or Injured? (Yes/No)

Stain Color?

____

Chemical Reagent used?
(Ammonia/$FeSO_4$/KOH)

Describe Reaction:________________________

____

Photos Taken? (Yes/No)

Spore Color?

____

Suspected ID:__________________________

Notes

____

____

____

REFERENCES & FURTHER READING

Akiyama, K., Matsuzaki, K. I., & Hayashi, H. (2005). Plant sesquiterpenes induce hyphal branching in arbuscular mycorrhizal fungi. *Nature*, *435*(7043), 824-827.

Alves, M. H., & do Nascimento, C. C. (2014). Mycena margarita (Murrill) Murrill, 1916 (Basidiomycota: Agaricales: Mycenaceae): a bioluminescent agaric first recorded in Brazil. *Check List*, *10*(1), 239-243.

Badham, E. R. (1984). Ethnobotany of psilocybin mushrooms, especially Psilocybe cubensis. *Journal of ethnopharmacology*, *10*(2), 249-254.

Banbury, G. H. (1955). Physiological studies in the Mucorales III. The zygotropism of zygophores of Mucor mucedo Brefeld. Journal of Experimental Botany, 6(2), 235-244.

Banik, M. T., Lindner, D. L., Ortiz-Santana, B., & Lodge, D. J. (2012). A new species of Laetiporus (Basidiomycota, Polyporales) from the Caribbean basin. Kurtziana, 37, 15-21.

Baroni, J. T. (n.d.) Species List: Belize. Retrieved from: http://facultyweb.cortland.edu/NeoTropicalFungi/Belize/list.asp

Baroni, T. J. (2017). *Mushrooms of the northeastern United States and eastern Canada*. Timber Press. 268

Barron, G. L. (2003). Predatory fungi, wood decay, and the carbon cycle. *Biodiversity*, 4(1), 3-9.

Bartnicki-Garcia, S. (1968). Cell wall chemistry, morphogenesis, and taxonomy of fungi. Annual Reviews in Microbiology, *22*(1), 87-108.

Bessette, A. (2007). Nature: Mushrooms of the Southeastern United States. *Syracuse University Press*

Beug, M. W. (2011). Pleurocybella porrigens toxin unmasked? *Journal of American Amateur Mycology*. Retrieved September 4 2016.

Billings, R. F., Clarke, S. R., Mendoza, V. E., Cabrera, P. G., Figueroa, B. M., Campos, J. R., & Baeza, G. (2004). Bark beetle outbreaks and fire: a devastating combination for Central America's pine forests. *Unasylva*. 217.

Bosmans, F., Maertens, C., Verdonck, F., & Tytgat, J. (2004). The poison Dart frog's batrachotoxin modulates Nav1. 8. *FEBS letters*, *577*(1-2), 245-248.

Boyd, P. E., & Kohlmeyer, J. (1982). The influence of temperature on the seasonal and geographic distribution of three marine fungi. *Mycologia*, 894-902.

Brundrett, M. C. (2002). Coevolution of roots and mycorrhizas of land plants.New phytologist, 154(2), 275-304.

Byrne, P. J., & Jones, E. G. (1975). Effect of salinity on the reproduction of terrestrial and marine fungi. *Transactions of the British Mycological Society*, *65*(2), 185-200.

Casselman, A. (2007). Strange but true: the largest organism on earth is a fungus. Sci Am, 4.

Curnow, P., & Tam, M. (2003). Contact dermatitis to Shiitake mushroom. *Australasian Journal of Dermatology*, *44*(2), 155-157.

De Borhegyi, S. F. (1961). Miniature mushroom stones from Guatemala. *American Antiquity*, *26*(4), 498-504.

De Hoog, G. S., Guarro, J., Gené, J., & Figueras, M. J. (2000). Atlas of clinical fungi (No. Ed. 2). Centraalbureau voor Schimmelcultures (CBS).

Desjardin, D. E., Perry, B. A., Lodge, D. J., Stevani, C. V., & Nagasawa, E. (2010). Luminescent Mycena: new and noteworthy species. *Mycologia*, *102*(2), 459-477.

Dourson, D. (2013). Creature Feature: Bio-luminescent Fungus. URL: http://www.bfreebz.org/2013/03/26/bioluminescing-fungus-glowing-belize/

Easton, R. M. (1994). Lichens and rocks: a review. Geoscience Canada,21(2).

Flemming, N. (2014, November). Plants talk to each other using an internet of fungus. *BBC*. Retrieved from BBC.com website http://www.bbc.com/earth/story/20141111-plants-have-a-hidden-internet

Grieve, M. (1971). A modern herbal: the medicinal, culinary, cosmetic and economic properties, cultivation and folk-lore of herbs, grasses, fungi, shrubs, & trees with all their modern scientific uses (Vol. 2). Courier Corporation.

Grob, C. S., Danforth, A. L., Chopra, G. S., Hagerty, M., McKay, C. R., Halberstadt, A. L., & Greer, G. R. (2011). Pilot study of psilocybin treatment for anxiety in patients with advanced-stage cancer. Archives of general psychiatry, *68*(1), 71-78.

Hass, H, Taylor, T.N. & Remy, W. (1994). Fungi from the Lower Devonian Rhynie chert: Mycoparasitism. *American Journal of Botany* 81(1):29-37.

Hibbett, D. S., Binder, M., Bischoff, J. F., Blackwell, M., Cannon, P. F., Eriksson, O. E., ... & Lumbsch, H. T. (2007). A higher-level phylogenetic classification of the Fungi. Mycological research, *111*(5), 509-547.

Hughes, D. P., Anderson, S.B., Hywel-Jones, N.L., Himaman, W., Billen, J. & Boomsma, J. J. (2011). Behavioral mechanisms and morphological symptoms of zombie ants dying from fungal infections. *BMC Ecology. 11*(13)

Hughes, W.O.H., Thomsen, L., Eilenberg, J. J., Boomsma. (2003). Diversity of entomopathogenic fungi near leaf-cutting ant nests in a neotropical forest, with particular reference to Metarhizium anisopliae var. anisopliae. *Journal of Invertebrate Pathology. 85*(1) 46-53

Husbands, D. R., Henkel, T. W., Bonito, G., Vilgalys, R., & Smith, M. E. (2013). New species of Xerocomus (Boletales) from the Guiana Shield, with notes on their mycorrhizal status and fruiting occurrence. *Mycologia*, *105*(2), 422-435.

Hyde, K. D., Jones, E. G., Leaño, E., Pointing, S. B., Poonyth, A. D., & Vrijmoed, L. L. (1998). Role of fungi in marine ecosystems. *Biodiversity & Conservation*, *7*(9), 1147-1161.

Isikhuemhen, O. S., Anoliefo, G. O., & Oghale, O. I. (2003). Bioremediation of crude oil polluted soil by the white rot fungus, *Pleurotus tuberregium* (Fr.) Sing. Environmental Science and Pollution Research, 10(2), 108-112.

James, T.Y., P.M. Letcher, J.E. Longcore, S.E. Mozley-Standridge, D. Porter, M.J. Powell, G.W. Griffith, and R. Vilgalys. 2006. A molecular phylogeny of the flagellated Fungi (Chytridiomycota) and description of a new phylum (Blastocladiomycota). *Mycologia* 98: 860-871.

Jones, E. G. (2000). Marine fungi: some factors influencing biodiversity. *Fungal Diversity*, *4*(193), 53-73.

Kirk, P. M., Cannon, P. F., Minter, D. W., & Stalpers, J. A. (2008). Dictionary of the Fungi. (10). Wallingford, UK.

Kirk, P.M., Cannon, P.F., David, J.C., and Stalpers, J. (2001). Ainsworth and Bisby's Dictionary of the Fungi. CAB International (9)

Kohlmeyer, J. (1969a). The role of marine fungi in the penetration of calcareous substances. *American Zoologist*, *9*(3), 741-746.

Kohlmeyer, J. (1969b). Deterioration of wood by marine fungi in the deep sea. In *Materials performance and the deep sea*. ASTM International.

Kohlmeyer, J. (1983). Geography of marine fungi. *Australian Journal of Botany Supplementary Series*, *13*(10), 67-76.
Kohlmeyer, J., & Kohlmeyer, E. (2013). *Marine mycology: the higher fungi*. Elsevier.
Kohlmeyer, J., Bebout, B., & Vlkmann-Kohlmeyer, B. (1995). Decomposition of mangrove wood by marine fungi and teredinids in Belize. *Marine Ecology*, *16*(1), 27-39.
Krogh, D (2010). Biology: A Guide to the Natural World, Benjamin-Cummings Publishing Company. 409
Kuo, M. (2006, November). Determining odor and taste. Retrieved from the MushroomExpert.Com Web site: http://www.mushroomexpert.com/odortaste.html
Kuo, M. (2006, November). Testing chemical reactions. Retrieved from the *MushroomExpert.Com* Web site: http://www.mushroomexpert.com/macrochemicals.html
Kuo, M. (2007, February). *Bolbitius reticulatus.* Retrieved from the *MushroomExpert.Com* Web site: http://www.mushroomexpert.com/bolbitius_reticulatus.html
Kuo, M. (2008, October). The genus *Xylaria*. Retrieved from the *MushroomExpert.Com* Web site: http://www.mushroomexpert.com/xylaria.html
Kuo, M. (2010, March). *Boletus variipes.* Retrieved from the *MushroomExpert.Com* Web site: http://www.mushroomexpert.com/boletus_variipes.html
Kuo, M. (2013, December). Cup fungi. Retrieved from the *MushroomExpert.Com* Web site: http://www.mushroomexpert.com/cups.html
Kuo, M. (2014, February). The bird's nest fungi. Retrieved from the MushroomExpert.Com Web site: http://www.mushroomexpert.com/birdsnests.html
Kuo, M. (2014, November). Waxy caps: The Hygrophoraceae family, in part. Retrieved from the *MushroomExpert.Com* Web site: http://www.mushroomexpert.com/hygrophoraceae.html
Kuo, M. (2017, November). *Boletellus ananas.* Retrieved from the *MushroomExpert.Com* Web site: http://www.mushroomexpert.com/boletellus_ananas.html
Liebeskind, A. (1972). Mycological problems in occupational allergies. *Journal of Asthma Research*, *10*(1), 71-73.
Maneevun, A., Dodgson, J., & Sanoamuang, N. (2012). Phaeoclavulina and Ramaria (Gomphaceae, Gomphales) from

Nam Nao National Park, Thailand. *Tropical Natural History*, *12*(2), 147-164.
Masuma, R., Yamaguchi, Y., Noumi, M., Omura, S., & Namikoshi, M. (2001). Effect of sea water concentration on hyphal growth and antimicrobial metabolite production in marine fungi. *Mycoscience*, *42*(5), 455-459.
McKnight, K. H., & McKnight, V. B. (1998). *A field guide to mushrooms: North America* (Vol. 34). Houghton Mifflin Harcourt.
Moreno-Rico, O., Sánchez-Martínez, G., Marmolejo-Monsiváis, J. G., Pérez-Hernández, K., & Moreno-Manzano, C. E. (2015). Diversidad de hongos Ophiostomatoides en pinos de la sierra Fría de Aguascalientes, México, asociados con Dendroctonus mexicanus. *Revista mexicana de biodiversidad*, *86*(1), 1-8.
Mueller, U. G., Rehner, S. A., & Schultz, T. R. (1998). The evolution of agriculture in ants. *Science*, *281*(5385), 2034-2038.
Nash, T. (2010). Lichen Biology. Cambridge University Press. (2)
O'Reilly, P. (2016) Fascinated by Fungi. *First Nature*.
Ortiz-Santana, B., Lodge, D. J., Baroni, T. J., & Both, E. E. (2007). Boletes from Belize and the Dominican republic. *Fungal Diversity*. 27: 247-416., 27.
Ovrebo, C. L., Lodge, D. J., & Aime, M. C. (2011). A new Cantharocybe from Belize with notes on the type of Cantharocybe gruberi. *Mycologia*, *103*(5), 1102-1109.
Petersen, R. H. (1967). Notes on clavarioid fungi. VII. Redefinition of the Clavaria vernalis-C. mucida complex. *American Midland Naturalist*, 205-221.
Poisson, J. (2010). "Raphaël Dubois, from pharmacy to bioluminescence". Rev Hist Pharm (Paris) (in French). France. 58 (365): 51–56.
Purtov, K. V., Petushkov, V. N., Baranov, M. S., Mineev, K. S., Rodionova, N. S., Kaskova, Z. M., ... & Medvedeva, S. E. (2015). The chemical basis of fungal bioluminescence. *Angewandte Chemie*, 127(28), 8242-8246.
Sanders, W. B. (2014). Complete life cycle of the lichen fungus Calopadia puiggarii (Pilocarpaceae, Ascomycetes) documented in situ: Propagule dispersal, establishment of symbiosis, thallus development, and formation of sexual and asexual reproductive structures. *American journal of botany*. 101(11), 1836-1848.

Schaefer, S. B., & Furst, P. T. (Eds.). (1997). *People of the peyote: Huichol Indian history, religion, and survival*. UNM Press.

Schultes, R.E. and Hofmann, A. (1980) The Botany and Chemistry of Hallucinogens, 2nd edn. Thomas, Springfield, IL.

Schüßler, A., & Walker, C. (2010). The Glomeromycota: a species list with new families and new genera.

Singer, R. (1958). Mycological investigations on teonanacatl, the Mexican hallucinogenic mushroom. Part I. The history of teonanacatl, field work and culture work. *Mycologia*, 239-261.

Swann, E and Hibbett D, S. (2007). Basidiomycota. The Club Fungi. Version 20 April 2007.http://tolweb.org/Basidiomycota/20520/2007.04.20 in The Tree of Life Web Project, http://tolweb.org/

T.J. Hodgson & W.K. Darrow. (2003). The Mountain Pine Ridge Reforestation Project, Belize CA. *XII World Forestry Congress*.

Thomas, O. R. (1979). *Mushrooms of Western North America*.

University of Mississippi. (2013). Marine Fungi: Species of Higher Marine Fungi. Retrieved from https://web.archive.org/web/20130422084649/http://ocean.otr.usm.edu/~w529014/index_files/Page2025.htm

Wainwright, M. (1992). An introduction to fungal biotechnology. *John Wiley & Sons*.

Wakimoto, T., Asakawa, T., Akahoshi, S., Suzuki, T., Nagai, K., Kawagishi, H., & Kan, T. (2011). Proof of the existence of an unstable amino acid: pleurocybellaziridine in Pleurocybella porrigens. *Angewandte Chemie International Edition*, *50*(5), 1168-1170.

Xudong, Z. H. O. U., De Beer, Z. W., Cibrian, D., Wingfield, B. D., & Wingfield, M. J. (2004). Characterisation of Ophiostoma species associated with pine bark beetles from Mexico, including O. pulvinisporum sp. nov. *Mycological Research*, *108*(6), 690-698.

Made in United States
Orlando, FL
05 November 2026

90241060R00100